T0282352

# Software Product Quality Control

Stefan Wagner

# Software Product Quality Control

Stefan Wagner
Institute of Software Technology
University of Stuttgart
Stuttgart
Germany

ISBN 978-3-642-44190-5 ISBN 978-3-642-38571-1 (eBook)
DOI 10.1007/978-3-642-38571-1
Springer Heidelberg New York Dordrecht London

ACM Computing Classification (1998): D.2, K.6


Softcover re-print of the Hardcover 1st edition 2013


Printed on acid-free paper

Springer is part of Springer Science+Business Media (www.springer.com)

*To Julia*

# Preface

This book has been a much longer process than I would have ever anticipated. The original idea was to integrate and combine the research on software product quality control with my then colleagues Florian Deissenboeck and Elmar Juergens, which we have done in close collaboration with industry to help practitioners in implementing quality control in practice. As life goes on, however, Florian and Elmar decided to start their own company, and, over time, it became clear that they cannot spend enough time on this book project. Hence, in 2011, I bravely decided to write the book on my own.

This led to some changes in the content, shifting away from areas I personally have not worked in so much, to other areas I could contribute more personal experience. In addition, I was the project leader for the consortium project Quamoco which had its focus on quality models and quality evaluation. The project and its result strongly influenced this book. I am very happy to be able to report on the results of this project which allowed me to integrate many things we have done before into a comprehensive approach.

I hope this book will be useful for practitioners, students and researchers interested in and working on software product quality assurance and quality control. I tried to be concise in the book so that it is possible to quickly understand all the concepts but at the same time give enough depth so that you can directly apply the techniques and approaches. In particular, I concentrated on reporting several practical experiences we have made with the techniques from the book hoping they can be models for other companies.

This book represents a summary of a lot of research that I have done over the last 10 years. Naturally, it is impossible to thank everybody who has contributed to this research in some way. I have to restrict myself to the ones directly contributing to what led to the contents of the book, and even then, I fear I will forget many who helped, supported and worked with me over the years. I thank Florian Deissenboeck and Elmar Juergens for starting this project with me and the lot of interesting research we have done together. I am grateful to Ivan Bogicevic, Martin Feilkas, Mario Gleirscher, Dimitriy Golubitskiy, Markus Herrmannsdoerfer, Benjamin Hummel, Maximilian Irlbeck, Klaus Lochmann, Daniel Méndez Fernández,

Daniel Kulesz, Markus Luckey, Holger Röder, Rainer Schmidberger, Sebastian Winter, Jinying Yu and all the members of the Quamoco project as well as our partners at the companies we have worked with. I am also grateful to the German Ministry for Education and Research which supported the Quamoco project (01IS08023B). I particularly thank Harry Sneed for his detailed feedback on an earlier version of this book. Finally, of course, I would like to thank my family for the long-term support, especially my mother Ottilie, my father Raimund and Julia.

Stuttgart, Germany
March 2013

Stefan Wagner

# Contents

# Chapter 1
# Introduction

## 1.1 Motivation

This chapter introduces and motivates software product quality control. It gives guidance how to read the book. Important terms are explained and discussed as basis for the further chapters.

All these web browsers, however, serve the same purpose: to display pages from the World Wide Web. They differ only slightly in the features they provide. In contrast, the non-functional properties of the browsers become more and more important. The users are interested in the quality of how the browsers can provide these features. The recent successes of Google's Chrome browser is at least partly due to its reputation of high performance. Good performance was also one of the reasons that Mozilla's Firefox initially gained many users. Another reason was that in the early 2000s, several security vulnerabilities in Microsoft's Internet Explorer 6 were published, which damaged its market position[1] and drove many users to alternative browsers.

Quality, however, is not a fixed and universal property of a software. It depends on the context and goals of its stakeholders. Hence, when we want to develop a high-quality software system, the first step should be a clear and precise specification of quality [26]. Even if we get the specification right and complete, we can be sure that it will become invalid over time, if we do not control quality over the complete life cycle of the software.

As Fig. 1.1 shows, software is clamped between the business and technical processes it has to support and the technical platform (hardware, operating system or database system) on which it runs. Both, the processes as well as the platform, will inevitably change over time. Hardware becomes obsolete, operating systems are upgraded to new versions, languages evolve, new tools are released, and business processes need to support new and changed businesses.

[1] http://en.wikipedia.org/wiki/Usage_share_of_web_browsers.

S. Wagner, *Software Product Quality Control*, DOI 10.1007/978-3-642-38571-1_1, 

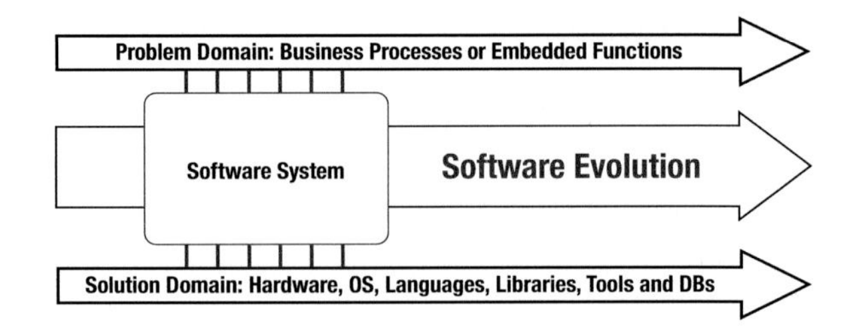


**Fig. 1.1** Software as mediator between problem and solution domain [46]

Hence, the needed quality of the software changes and if the software does not change accordingly, its quality will decay. The software ages [170]. The most problematic part of it is that "By the time you figure out you have a quality problem it is probably too late to fix it" [182]. I believe a major reason for these problems is that software is intangible. In contrast to other engineering disciplines, we have no direct visual and haptic feedback on the progress and state of our products. We cannot take a software prototype into our hands and play with it or just see any damages in the material. We either see an uncountable number of screens full of text or an execution of the software which may or may not have a graphical user interface. To some degree, many modern engineering disciplines have that problem: Also mechanical systems are nowadays designed and simulated virtually on a computer. In software engineering, however, anything stays virtual over the whole life cycle of the product.

Hence, it takes considerable effort to get a good picture of the current state of a software system. We need to make various analyses and interpret the results in the appropriate context of that particular software system. Therefore, we are specifically prone to ignore the real state of a system and, thereby, quality problems. This might be an instance of the psychological phenomenon called *cognitive dissonance* [63]. It describes the discomfort we experience when we hold "conflicting cognitions simultaneously". In this case, we build a software system with our best effort and are therefore confident that it will have a high quality. At the same time, we see problems with the system. For example, it might be slow in first tests, we have problems understanding old code, or the changes tend to require changes in many parts of the system. Psychology says that people then tend to reduce the importance of one of the cognitions to create a consistency between the conflicting beliefs. Here, we tend to ignore the signs of bad quality in favour of our belief that we make high-quality work.

The only solution is continuous quality control: the steady and explicit evaluation of a product's properties with respect to its quality goals. You have probably heard of *continuous integration* which became well known as one of the main practices of Extreme Programming (XP) [13]. The idea of continuous integration is to integrate often, at least daily, to have small integration increments, detect integration problems early and remove them easily. The alternative is a large integration at the end of the implementation what often results in many, hard-to-manage problems. Practitioners like to refer to that as "integration hell". Instead, with continuous integration, the integration becomes a seamless part of normal development.

Continuous integration includes already to build and run automated tests. With continuous quality control, we go beyond that: We reassess every day if our quality goals are still aligned with our business goals, if our software fits the platform, if it is equipped for future changes and if it fulfils our quality goals. We synthesise our quality goals into quality models and perform various analyses, tests and reviews; integrate the results; and identify quality problems. The analysis results and quality problems are made visible and available to all engineers to generate a joint effort to remove problems and avoid them in the future. This book will show you a way how you can achieve that in your development and maintenance projects.

I have to stress, however, that this book is not about controlling people. It is about taking control of the situation, control of the project and control of the product. As project leader, quality manager or developer, you need to have the best possible information about your product to make good decisions. This may sound obvious, but how many software projects do you know in which the team members know the amount of cloning in their source code, the module with the most warnings from static analysis tools or even the exact number for the size of the source code? We have to work actively to collect and update this information.

Continuous quality control needs an investment to set up a functioning product quality control system, but it pays back in a short time by concentrating efforts to the right tasks and avoiding costly defects in late project phases. This puts us as software engineers back in control and allows us to develop the best possible product. I will present in this book (1) the results of the three-year project *Quamoco* of a German consortium which built an operationalised quality model that can be used as a basis for quality control and (2) our experiences with various aspects of quality control in practice.

## 1.2 How to Read This Book

The aim of this book is to guide and support you in setting up and running continuous quality control in your environment. Therefore, it is mainly oriented at practitioners. It contains enough fundamental and theoretical contents, however, that it should be useful to software engineering students in a software quality or quality assurance class. The contents of the book are integrated but the main chapters should be readable also when you have not read the preceding chapters completely. So feel free to jump from one section you are interested in to others. If I build on other parts of the book, there are always references to follow. I also paid special attention to summarising the contents in easily accessible checklists and to referring to further reading, so you can dig more deeply in the areas you want to know more about.

The intended audience in general are all people in software engineering and software management interested in modern software product quality techniques and processes. As mentioned, I mostly aim at practitioners, especially software engineers, responsible for the quality assurance processes in a company. They probably can benefit most from reading this book by getting stimuli to improve their current processes (or get confirmation that their current processes are good). Also software engineers mainly concerned with programming will profit from this book, because they will learn about quality control techniques and many of those can also be applied by single programmers. Software managers can learn about how they can get a good picture of the quality of their systems and get help in planning it. Finally, students in computer science, software engineering and related subjects can use this book in practice-oriented courses on software quality, quality assurance and maintenance.

This book was written in a way so that you can read it from introduction to summary to guide you from general concepts over detailed information to practical experiences and conclusions. You are reading Chap. 1 right now, in which I motivate and introduce the topic of continuous quality control and define many terms we will use in the latter chapters. In Chap. 2, we discuss the foundation of quality control: quality models. We will see the historical development of quality models and a modern approach to quality modelling using the results of the research project Quamoco.

A good part of software engineering consists of planning. Also in quality control, we need to plan the desired qualities of our products as well as the quality assurance to ensure these qualities. I will present approaches for that in Chap. 3. As good approaches in this area are scarce, this is one of the shortest chapters. The longest chapter is Chap. 4. It describes the main concepts and techniques of continuous quality control. We will discuss the quality control loop as well as the basics and empirical knowledge about the main techniques used there such as reviews or testing.

Although I included examples all along these chapters, it was important for me to include separate practical experiences we have made with these approaches and techniques in Chap. 5. It contains several applications of different subsets of the presented quality control techniques with real companies with a special focus on what we have learned. Finally, I summarise the main concepts in Chap. 6 and give more general further readings on the topics of this book.

I depicted four general ways through this book in Fig. 1.2 depending on your focus of interest. As mentioned before, the book was written so that you can read start to finish and the chapters will build on each other. This is called *Complete* in the figure. If you mainly want to learn about modern software quality models, you can follow the *Quality Model Focus* which goes from the introduction directly to the quality models and finishes with the corresponding parts in the practical experiences. If you are more interested in the control loop and the corresponding quality assurance techniques, you can follow the *Quality Control Focus* which continues after the introduction with the chapter on quality control and the corresponding practical experiences. I suggest, however, that at some point

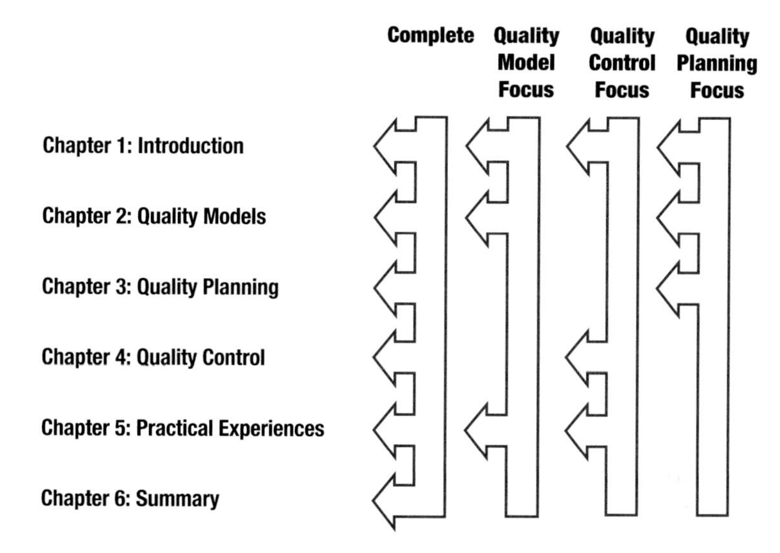


**Fig. 1.2** Four possible ways through the book

you also look into quality models as they can greatly help to improve and manage quality control. Finally, if you are looking for support in quality planning, you can use the *Quality Planning Focus*, which concentrates on the first three chapters of the book. I would advise to include the quality model chapter because we rely strongly on them in the quality planning part of the book.

I tried to make this book compact so you can get a concise but complete impression of software quality control. As the area of software quality is huge, this also means that this book cannot be exhaustive. For some topics, which I consider important but which were not necessary for the main flow of the book, I included basic information in *sidebars*. Each sidebar introduces the corresponding topic and guides you to further information. To help you in transferring the techniques from this book into practice, I included checklists where appropriate which condense the information for quick reference. Finally, I also included further readings at many points in the book to help you in getting more information.

## 1.3 Software Quality

What is *software quality*? Or even, what is *quality*? Quality is a concept that has kept philosophers occupied for more than 2,000 years. The roots of describing and defining quality probably lie in the view of Plato on beauty [173]. He believed that beauty causes responses of love and desire but that beauty is located in the beautiful thing. Hence, we can objectively decide if something is beautiful or not. Over time, philosophers moved to a contrary view, in which beauty is what causes desire. This even led to a view in which beauty means something is useful.

Most of this discussion can easily be transferred to our modern use of the word *quality*. Originally it did not contain this notion of an evaluation as good or bad, but in contemporary usage, we usually mean that a quality software is a good software. For this evaluation, however, the whole century-long philosophical discussion applies: Is quality objectively contained in the software? Or does quality always depend on its usefulness for a specific user?

The ISO, IEC and IEEE define *quality* in [108] with six alternatives:

1. The degree to which a system, component or process meets specified requirements
2. The ability of a product, service, system, component or process to meet customer or user needs, expectations or requirements
3. The totality of characteristics of an entity that bear on its ability to satisfy stated and implied needs
4. Conformity to user expectations, conformity to user requirements, customer satisfaction, reliability and level of defects present
5. The degree to which a set of inherent characteristics fulfils requirements
6. The degree to which a system, component or process meets customer or user needs or expectations

This variety reflects the slightly different meanings of quality as used in international software standards. In most definitions, we find the notion of requirements that need to be satisfied to have quality. These requirements, and the corresponding quality, can be on a system, component, process, product or service. Hence, we can talk about the quality of all of these entities. We will especially discuss the differences in product quality and process quality below (Sect. 1.3.2). Furthermore, the requirements to be satisfied can come from the user or the customer. So it is not clear if high quality means we satisfy what the person using the system or what the person paying for it wants it to do. What is even worse is that requirements can be implicit or only user expectations. Hence, even if we fulfil all explicitly specified requirements, our system might not be of high quality because it does not fit to user expectations. Even with a very elaborate and detailed requirements analysis, it is still possible to miss important requirements in a specification. So what do we need to look for in quality?

### *1.3.1 Garvin's Quality Approaches*

This diversity in quality definitions is not unique to software quality. Garvin [71] set out to give a comprehensive understanding of *product quality* by defining a set of different views or approaches to quality, which we can transfer to software quality [122]:

- Transcendent approach
- Product-based approach

- User-based approach
- Manufacturing approach
- Value-based approach

The *transcendent* approach is, as its name suggests, the most diffuse, hard-to-concretise view on product quality. It captures the intuitive feeling that a product has a high quality. We often see this view in requirements engineering in statements from customers who have a hard time specifying what and with what quality they want it, such as "I know it when I see it". This approach is closest to Plato's view of beauty as an inherent and undefinable property. It helps us little in software product quality control but to recognise that we cannot objectively measure everything that influences one's impression of quality.

A step more concrete is the *user-based* approach. It assumes that the product that satisfies the needs of the user best has the highest quality. The emphasis is not on specified user requirements, however, but on the subjective impression of the users. Garvin also discusses that user satisfaction and quality are not necessarily the same. The reason is that users can be satisfied with a product they do not consider of high quality. For example, a product can be not as well produced as high-quality products but is cheap enough to leave the customer satisfied. The user-based view contains the software quality definitions that mention implied user requirements or user expectations. For software quality control, this means that we need to include the user expectations into analysing and evaluating a software product.

The *value-based* approach is close to the user-based approach as the last example has shown. A cheap but less durable product can be of higher total value to a user than a durable but very expensive product. In this approach, we assign costs to conformance and nonconformance to requirements, compare it to the benefits of the product and, hence, can calculate its value. We assumes that we are able to assign a value to all involved factors. This actually blends two concepts as Garvin points out: "Quality, which is a measure of excellence, is being equated with value, which is a measure of worth". This can be useful in some cases in software quality control. We will discuss quality costs in detail below in Sect. 1.3.4.

In the *product-based* approach, quality describes differences in the quantity of some desired attributes of the product. Hence, in sharp contrast to the transcendent approach, this is precisely measurable. We assume that we exactly know and are able to describe what is desired. For example, the quality of tyres can be measured with the time they can safely be used. This approach is difficult for software, however, because such metrics either do not exist for certain qualities or are very hard to measure. For example, the maintainability of a software could be determined by the effort needed to complete a change request. The effort for a change request is not easy to measure because developers usually do not keep detailed and precisely track of what they are working on every minute. Furthermore, the effort also depends on the complexity of the change requests. Hence, we would have to normalise the effort by this complexity which is also difficult to measure. There are certain measurements we can perform, however, to get a more objective evaluation of a software's quality.

Note that the easy-to-measure and often used quantities *number of found defects* or *defects per kLOC* are not a *desired* attribute of a software product. In the product-based approach, we only consider "ideal" attributes. Such metrics are useful, however, in the *manufacturing* approach, which takes a more internal view and defines quality as conformance with specified requirements. This definition, however, includes all the problems we have with developing exhaustive and useful specifications. It assumes that it is always possible to define the requirements of a product completely and, hence, a deviation of the specification can be easily recognised. The metric *defects per kLOC* from above is then useful as we measure the deviation from the specification by the defect density of the code. In software quality control, a more appropriate name for this approach would be the *process* approach, because we do not have manufacturing of software but a development process. Hence, we also have to ensure that the development process creates the software in a way that conforms to its specification.

How can we now apply these approaches? Garvin suggests that different approaches are more relevant at different points in time in the product's life cycle. In the beginning of the life cycle, the product's inception, we need to focus on the user-based and value-based approaches to understand what is most important and most valuable for our users and customers. We could call this product and quality *goals*. Afterwards, we need to refine these goals to more concrete product attributes in the product-based approach. In other words, we create a specification of the product. Finally, when we build the product, we focus on the manufacturing approach which ensures that we build the product suitable for the specification.

### *1.3.2 Product Quality vs. Process Quality*

I already emphasised in the title of this book that we will focus on *product quality* which I especially emphasise as counterpart to *process quality*. Most of the research over the last twenty years has concentrated on investigating process quality. In some sense, this is the manufacturing approach of Garvin but not exactly. Garvin stressed the conformance to requirements, while in the common notion of *process quality* attributes of the process are evaluated. The idea is, as in other, manufacturing-heavy industries, high-quality processes also lead to high-quality products [192].

A widely used standard with this manufacturing and process view is ISO 9000 [90]. It contains the idea of establishing a quality management system in a company so that the resulting quality of the product will also be high. It does not concern itself with the quality of the products, however, but only with quality requirements on the company producing the products. Especially in the 1990s, a creation of such a quality management system was very popular. A factor in this might be that it is possible to be certified with ISO 9000 and its related standards which can be used as a marketing argument. Besides marketing, however, introducing ISO 9000 can have many benefits for a software company, such as

**Table 1.1** Delivered defects per thousand function point at CMM levels (based on [110])

| CMM level | Minimum | Average | Maximum |
|---|---|---|---|
| 1 | 150 | 750 | 4,500 |
| 2 | 120 | 624 | 3,600 |
| 3 | 75 | 473 | 2,250 |
| 4 | 23 | 228 | 1,200 |
| 5 | 2 | 105 | 500 |

making quality assurance processes explicit and clear. It also comes at the cost of higher bureaucracy and a lot of paperwork.

In the same spirit, two initiatives created standards particularly to define and improve software development processes: CMMI and SPICE. CMMI [40] was initially an American standard but is accepted worldwide today. The European SPICE has become the ISO standard ISO 15504 [93] and is also in use worldwide. Both standards rely on the idea of a normative, prescriptive approach to process improvement. In other words, there is an ideal process, described in the standards, that every company needs to achieve. There are several maturity or capability levels on the way from chaotic processes to highly standardised and optimised processes. An attractive feature of these standards is also that it is possible to be certified at a certain level of maturity. Again, besides certifications, this in-depth analysis and improvement of processes helps in software quality but comes at the price of more inflexible and more bureaucratic process.

Another problem is that this process orientation leads to "quality assessment that is virtually independent of the product itself" [122]. Hence, we rely solely on the assumption that good processes produce good software. This clear relationship between process and product quality may be well established for manufacturing. For development processes, it is far less clear [122, 199]. For example, Jones [110] investigated the relationship between the CMM (the predecessor of CMMI) level of a company and the number of shipped defects per function point in their products. As a result, he saw on average that the number of defects decreases with rising, i.e. better, CMM level as shown in Table 1.1. Therefore, there is a relationship between process and product. The problem is, however, that on the extremes, the best companies at CMM level 1 (worst) produce software with less defects than the worst companies at level 5 (best). Hence, there are clearly more factors influencing the defect rate.

In conclusion, I would like to emphasise that processes and the quality of processes are important to deliver high-quality software products. Nevertheless, many factors influence product quality and, therefore, we need to evaluate and monitor product quality directly as well as improve the processes that create these products. As there are many books on ISO 9000, CMMI and SPICE, we will concentrate on product quality in the following.

### 1.3.3 Product Quality

Our view of product quality in this book is the combination of Garvin's approaches to product quality described above: We need to understand the user expectations, translate them into clear product attributes and ensure that these attributes are then implemented in the product. To better understand and discuss these user expectations and product attributes, we will explore in more detail what different areas of software product quality we usually have to consider.

Such a taxonomy of software quality has been subject to research for several decades. The result is a series of so-called *quality models*. We will discuss them in detail in Chap. 2. The main idea of most quality models is to break down the complex concept "quality" in *quality factors* that may be broken down again so that we get a hierarchy of quality concepts. This is useful in many ways but the main and most direct use is as a checklist during requirements analysis. Examples for quality factors are *security* and *maintainability*. Therefore, instead of asking only about the expected features and quality, we can discuss with the stakeholders how secure and how maintainable they want the software to be. This is still abstract but nevertheless easier to discuss. A collection of such quality factors in a quality model helps us not to forget an aspect of software quality. Furthermore, we can use it to derive different ways of measuring and evaluating quality factors to come to a comprehensive quality evaluation.

The most well-known taxonomy of software quality is the quality model in ISO/IEC 9126 [107], now superseded by the new standard ISO/IEC 25010 [97]. You will come across these standards at several places in this book. For now, we will only take the highest-level quality factors from ISO/IEC 25010 to discuss a common decomposition of software quality. We will discuss each of these quality factors in detail in the following:

- Functional suitability
- Reliability
- Performance efficiency
- Usability
- Security
- Maintainability
- Portability
- Compatibility

As quality is achieving user requirements and expectations, in a strict sense, it includes also the required functionality. This is reflected by the quality factor *functional suitability*, which expresses that the software shall provide the functionality to the user that fits to their requirements and expectations. This also includes functional *correctness*, i.e. that the software does what is required. In many contexts, correctness is equated with quality. It is only one specific aspect, however.

The other quality factor often equated with quality is *reliability*. It describes how frequently the software does not provide the expected or specified service.

Depending on your notion of providing a service, this can include most of the other quality factors. In general, however, we consider failures in the functionality of the software to distinguish it from performance problems, for example. Hence, reliability is a comparably well-defined quality factor for which statistical models and measurements exist. There is also a relationship to *correctness*: A correct software should not fail and, hence, be reliable. The relationship is more complex, however, as the environment in which the software is executed plays a large role if the system fails or not.

A quality factor most software engineers are well aware of is *performance efficiency*, often only called *performance*. It describes how efficiently the hardware resources are used by the software and in what time the users get a response from the software. There is a large body of well-founded theories on complexity that influences performance. There are various other factors influencing performance as well, which are not as well understood. We capture all that in this quality factor.

Another quality factor, which could subsume almost all other quality factors, is *usability*. It describes how well and with what satisfaction a user can operate the software. Hence, functional suitability, reliability and performance all play a role in usability. If the software does not offer the functionality I need, it is less usable for me. If it frequently crashes, it is less usable for me. If it reacts slowly, it is less usable for me. Therefore, usability is strongly connected to these factors. It has its specific focus on how well the users can perform their tasks, however. It is very close to Garvin's user-based approach ignoring specifications but concentrating on the user experience.

*Security* was not a top-level quality factor in ISO/IEC 9126 but has become so important that this was changed in ISO/IEC 25010. It describes how well prepared the software is against attacks from malicious attackers. Therefore, we leave the area of unwanted mistakes that lead to unsatisfactory behaviour of the software and go into harming the software on purpose. This has always been possible in computer systems but is now relevant for almost any software, because it can potentially be on a computer with connection to the Internet. The protection of the data and service of our software always also involves the hardware and further environment and is, hence, a quality factor for the whole system, not only software.

So far, all quality factors have aimed at the user. The following three have at least partially the view of the developers of the software. Clearly focused on the developers is *maintainability*. The term *maintenance* for all phases after the initial development is misleading as we do not have to change parts of a software because of wear and tear. We need to change the software so that it continues to fit into its environment and to improve it. Therefore, maintenance is essentially further development. Besides the term, it includes how easy it is to understand, change and extend the software. In some contexts, this is also called *code quality* or *internal quality*.

In case we want to bring our software to a new or further platforms, *portability* is important. It expresses how strongly tied the software is to the platform. How important this is for a product is differing strongly. A software product might be built for a very specific platform with no intention of using it on other platforms.

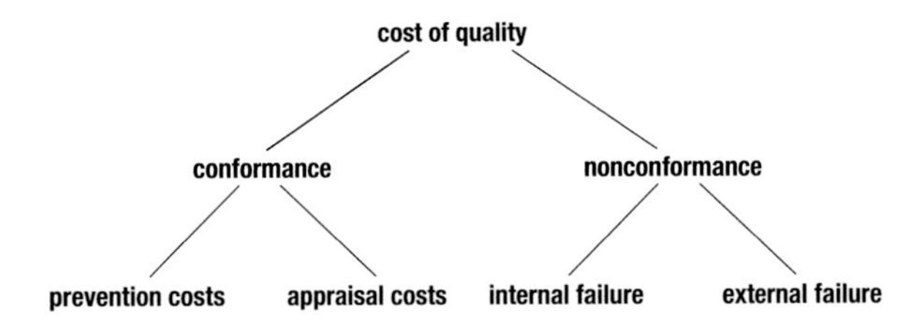


**Fig. 1.3** Cost types

Then a complicated architecture abstracting from the platform is probably a bad investment. In other contexts, nowadays mobile applications, it is probable that we want to port the software from one platform to others if it is successful.

Finally, *compatibility* is somewhere between the needs of users and developers. It can mean that a user can easily combine the software with other software and hardware systems, for example, that a mobile application is compatible with a cloud service to store personal data. It can also mean that developers can change components or services with little effort because there is a clear interface, maybe even a standardised API. Both issues become more and more interesting in current contexts in which many cloud and other Internet services become available.

### *1.3.4 Cost of Quality*

How can we now analyse, measure and evaluate all these different quality factors in a comprehensive way? How can we sum a usability problem with a maintainability problem? One unifying way to talk about quality is to express the various factors in monetary units. Stemming from the area of manufacturing, we talk about the *cost of quality* which describes the cost of quality improvement as well as the cost of missing quality. Quality cost models describe the different types of costs and their relationships.

The cost of quality is an area that is under research in various domains, more recently also in non-manufacturing areas. It describes the costs that are associated with preventing, finding and correcting defective work. These costs are divided into *conformance* and *nonconformance* costs. They are also called *control costs* and *failure of control costs*, which fits nicely to the topic of this book: What costs are necessary for controlling quality and what costs incur if we fail to control quality? We can further break down the costs into the distinction between prevention, appraisal and failure costs which gives the model the name *PAF model* [114]. The basic model was derived from the manufacturing industry but has been used repeatedly for software quality as well [126, 131, 190]. You can find a depiction in Fig. 1.3.

The *conformance costs* comprise all costs that need to be spent to build and control the software in a way that it conforms to its quality requirements. We further break this down to *prevention* and *appraisal* costs. Prevention costs are, for example, developer training, tool costs or quality audits, i.e. costs for means to prevent the

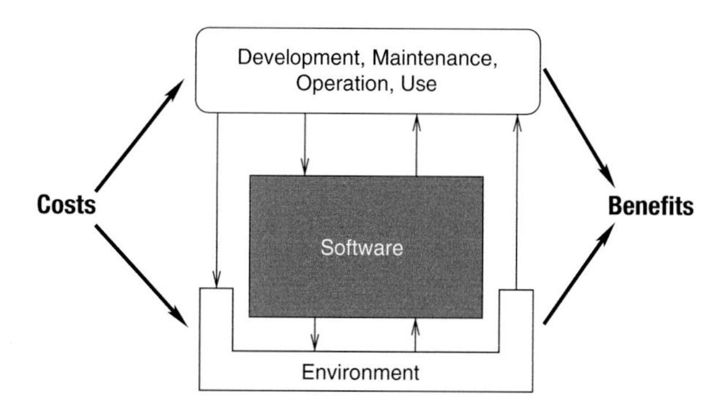


**Fig. 1.4** The relationship of quality and economics [201]

injection of faults. Appraisal costs incur by performing quality control techniques such as tests and reviews.

The *nonconformance costs* incur when the software does not conform to its quality requirements. These costs consist of *internal failure* costs and *external failure* costs. The former contain costs caused by failures that occurred in-house in the development project and the latter describe costs that result from failures after delivery to the users. The term *failure* is not to be interpreted exactly the same as we will define it in Sect. 1.4. A failure means that something went wrong in the development and, hence, appraisal costs can also describe the costs for reviewing source code. The PAF model for the costs of quality is a widely accepted basis for software quality economics. It supports primarily the manufacturing approach to quality, i.e. focuses on conformance and non-conformance to the specification or implied requirements. The problem with this model is that it stays rather abstract and it needs to be refined to be used in practice. Furthermore, existing work on software quality costs is mostly focused on the quality factor *reliability* [22, 85, 141]. Other quality factors, especially developer-oriented quality factors such as maintainability, are usually not considered.

Nevertheless, software quality is inseparably connected to its influence on the economics, i.e. the cost/benefit relation, of the software. Figure 1.4 shows this connection graphically. Each software is embedded in some environment consisting of other software components, platform software and hardware. There are also various activities that are performed on the software:

- The initial *development* of the software.
- *Maintenance* of the software consisting of corrective, adaptive and perfective changes.
- The software needs to be administrated during *operation*.
- The primary purpose is that the software is in *use* performing its function.

These two blocks – the environment and the activities – incur costs and generate benefits in combination with the software. The benefits are often saved costs but can also be independent of earlier costs. For example, the benefits can be shorter production times or new overseas markets because of online marketing. The environment has also an influence on the activities. It depends on whether we consider the software (e.g. an enterprise resource planning system) or the combined hardware/software system (e.g. an airbag control system) as the main product. In the former case, the environment can be seen as an addition to the software that can be cheaper or more expensive depending on the demands of the software. In the latter case we should handle the environment equally as the software and consider the influences on the activities directly.

The quality of the software is then how it influences the activities and its environment in incurring costs and generating benefits. In Fig. 1.4, the thin arrows represent the quality of the software (and its environment). Only these influences allow to judge whether a software is of "good" quality. This again shows that quality is a multifaceted concept that lies strongly in the eye of the beholder. If there is this tight relation between quality and economics, we do we not use quality economics models to evaluate software quality? There are at least five problems with this:

### Consideration of Economics in Quality Models

The available models of software quality aim at decomposing quality along one dimension. Activities and properties of the system are intermingled in a single dimension. This neglects this relationship of quality and economics. Moreover, the existing quality models tend to suggest that quality is an intrinsic property of the software although – as pointed out above – mainly the influences on the activities performed on the software determine the quality. One solution to address this is to include activities in the quality model. We will discuss that in Sect. 2.4.7.

### Empirical Knowledge in Research

Empirical research in software engineering is well established by now but it has still not gained the importance it deserves. For an economic analysis of software quality, most questions can only be answered empirically. Some questions have been worked on by the empirical software engineering community such as the efficiency of testing and inspection techniques, e.g. [9,115]. These issues are now better understood. The complete relation to economics, however, is still not clear.

Moreover, there are other factors that influence software quality and they need to be identified by field studies and experiments. Another problem is that studies at companies are difficult because the cost data is often considered as very sensitive, i.e. secret information that must not be given to competitors. Only then a better comparability of the own efforts with the domain average is possible.

### Data in Industry

The main fact that hampers analyses of quality economics in practice is the little data that is available in industry. There are several reasons for that. The main reason is the effort needed for collecting appropriate data. It is often the case that companies might be interested in results about quality and economic analyses but are not willing to invest the necessary effort. So far, we do not know what the cost/benefit relation of such a measurement program would be.

#### Monetary Units

I argued for economics because money is the only possible unit that is universal enough to measure the influences on quality. The monetary units are not as universal as I first suggested, however. The value of money changes over time, with different currencies and over different locations. The main problem is inflation. How can we build empirical knowledge using a continuously changing unit? One answer is to use net values and other methods that allow to inflation adjust the values. However, this renders meta-analysis even more difficult as it is today (cf. [200]).

On top of this, we have to deal with different currencies that have also ever-changing exchange rates. This also hampers comparability. A solution would be to convert all data into one broadly accepted currency. Moreover, the labour rates can differ even in a single currency. Because human labour costs constitute the main costs in software development, this complicates comparisons strongly.

#### Education

Computer science education is largely governed by its root in formal mathematics. Software engineering reality also includes economics but is not accurately represented. Very basic techniques such as statistical process control or building stochastic models are scarcely taught. I believe basic economics as well as quantitative and qualitative management would be an improvement in many curricula. This way, graduates would have a better understanding of the relationships between software quality and economics. This is often difficult to realise at the universities because also economists are usually not familiar with software engineering and have difficulties with teaching economics in a way useful for software engineers.

Economics and quality are two fundamental concepts in software development that are closely related. A thorough management of one of the two implies also the consideration of the other. There are several problems in research, education and practice today that hamper the modelling and evaluation of these relations. I will not be able to present a general solution to these problems in this book, but as pointed out before, the later covered activity-based quality models are one way to include these influences in quality modelling.

### *1.3.5 Dependable Software Systems*

Any kind of software product has quality requirements. It just differs how comprehensive and how strong they are. A throw-away prototype probably has only very

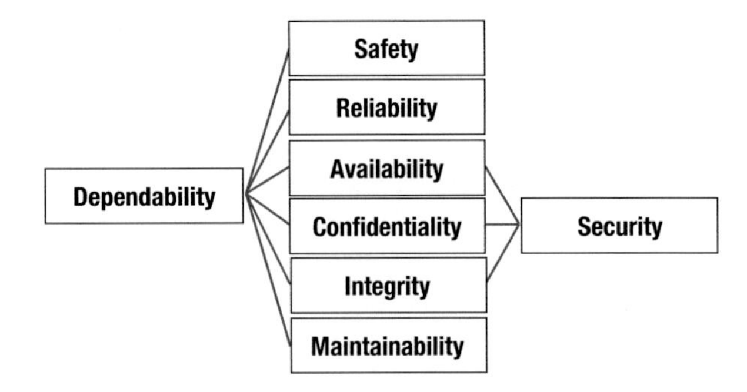


**Fig. 1.5** Dependability model of Avižienis et al. [4]

few important qualities, most often the functional suitability of the functionality that should be tried with it. On the other end of the spectrum, there is software with very high and very strong quality requirements, for example, embedded software systems controlling airbags in cars or business software performing bank transactions. This kind of software is often subsumed under the name *dependable software systems*.

This led to a dependability classification – or dependability model – by Avižienis et al. [4] shown in Fig. 1.5. The used quality factors are similar to the ones we saw in the ISO/IEC 25010 quality model above (Sect. 1.3.3). When we look at the subfactors in ISO/IEC 25010, we actually find all the quality factors apart from *dependability*. Only the arrangement is different. *Availability* is a part of *reliability* and not *security*. Furthermore, we have not seen *safety* before. It is mentioned in another quality model in ISO/IEC 25010, in the *quality in use* model, where it is a part of *freedom from risk*. This is the problem with taxonomies: They are never unique. The basic concepts are similar, however.

We notice that security is a big part of dependability. Hence, dependable systems should be secure. Security explicitly contains availability because certain kinds of attacks, such as *denial of service*, specifically aim at bringing the system down and making it not available to the users. This is very problematic, for example, if you lose business when your system is down. Customers will buy somewhere else when a web shop is not reachable. This, in turn, is closely related to the concept of reliability. Availability says for a defined time range or number of requests, how often it will be available, for example 99.9 %. This can be calculated out of the reliability, which says how probable it is that the software will not fail in a given time period. This is again an example of strongly interrelated quality factors.

We already know the quality factor *reliability*. It is an important factor for virtually any software product, but there is software that needs to be highly or *ultrareliable*. This is usually the case for software where any failure or a small amount of failure on demand already has dramatic consequences. This can be highly frequented websites as well as flight control systems. Such kind of software needs exceptional investments in quality control, such as formal methods (see below) and intensive testing.

Not to be confused with reliability is *safety*. Safety denotes the degree to which the system does not harm its environment, in particular humans but also valuable assets or the environment. A reliable software can contribute to a safe system but it can never make it safe. Sometimes, low reliability can be safer than high reliability. A plane on the ground is not reliable as it does not provide its intended service but it is quite safe as it cannot crash on the ground. In addition, it does not make sense to talk about software safety in an isolated way. Software can never be unsafe in itself, because it cannot harm someone directly. It always has to be embedded in a system containing hardware and mechanics which can create hazards. Hence, this is more a systems engineering topic. We will not go into more details of building safe software here.

As discussed above, *security* is becoming more and more important for any kind of software, but for dependable software, with high requirements on reliability and safety, any successful attack can have dramatic consequences. Hence, protection mechanisms and corresponding reviews and penetration testing are important for quality control.

*Maintainability* is also a general topic for software of any kind. In dependable software systems, it is sometimes stressed that *repairability* or *recoverability* is most important. This means that the software can quickly be fixed and it can come up and running again in a short time. This has effects on reliability. The downtime influences the time period a system provides its intended service. Longer downtime leads to lower reliability and availability.

For dependable systems, *formal methods* slowly start to play a role. With formal methods, we usually mean well-founded, mathematical methods that allow us to explore the state space of the software and prove important properties. Two important methods are *model checking* and *interactive theorem proving*. Especially model checking promises a high level of automation, which could be interesting to engineers who might not be trained in theorem proving. There are still many problems, however: The state space explosion in model checking allows, so far, only the analysis of small models. This means that we need to introduce strong abstractions in our models to be able to analyse them. The question then is if the abstractions are correct. Furthermore, in model checking, we need to formulate the properties we want to check in temporal logic, which is not familiar to most engineers. In conclusion, this area is interesting and could help in complex, dependable systems but is still limited for practical application in quality control today.

### 1.3.6 Quality Changes Over Time

A lot of the software we have today lives very long. There is also short-lived software such as very specific mobile applications or web applications. Most software, however, lives longer that the developers expect in the beginning. The Y2K problem, that software only used two digits to store the year and, therefore, was not able to calculate correctly after the year 2000, is a good example of that.

The developers of the Cobol systems in the 1970s did not expect their software to be still in use after 30 years, but it still is and it probably will be in the future.

Although software is intangible and, hence, not subject to wear and tear, its quality changes over time. As we have discussed above, the environment of the software changes: programming languages and platforms change. The business requirements change. The software gets bug fixes and new features. Hence, as long as we do not actively work against it, the quality of our software will decrease over time because of external and internal pressure.

Lehman [134] introduced a classification of the complexity of software systems, for example, with S-type programs which are small and completely specifiable. That is not the case for most of the software with which we work. Most of our current software will be E-type programs. This kind of programs is embedded in larger systems with which they have feedback. These larger systems could be electric/electronic systems, social systems or business systems. Lehman argues that the E-type programs have a cyclic dependency to this reality they are embedded in and, thereby, there cannot be a stable specification for these systems. We continuously need to re-evaluate our requirements, design and implementation.

Lehman summarised this also in (this selection of) his laws of software evolution [16]:

> I. Law of continuing change: *E*-type systems must be continually adopted else they become progressively less satisfactory.
> II. Law of increasing complexity: As an *E*-type system evolves its complexity increases unless work is done to maintain or reduce it.
> VII. Law of declining quality: The quality of *E*-type systems will appear to be declining unless they are rigorously maintained and adapted to operational environment changes.

These laws have since been confirmed in various empirical studies showing the size increase [70] or the quality decay, for example, in the architecture structure [56, 80, 117]. Parnas [170] formulated a similar observation in his concept of *software ageing*. Although a program as such does not age in the sense that because of time going by it deteriorates, it has to follow the external factors of its environment and, hence, needs to be changed. If no counteractions are taken, this will lead to ageing, because of what Parnas calls "ignorant surgery": Changes are introduced without understanding the original design. Continuous quality control will help you in preventing the most hurting signs of ageing.

## 1.4 Terms and Definitions

The previous sections showed that not only quality is a concept difficult to grasp but also the terminology in this context is vague and ambiguous. To address that, this section introduces the terms most relevant for this book with clearly defined and easy-to-understand definitions. Already *software product quality control*, the name of this book, has no standardised meaning yet [108]. Therefore, I give here my own definition based on [46, 108]:

**Definition 1.1 (Software product quality control).** A process to specify quality requirements, evaluate the created artefacts, compare the desired with the actual quality and take necessary actions to correct differences.

### *1.4.1 Quality Assurance*

A term very closely related and in many definitions not clearly distinguishable from quality control is *quality assurance*. In this book, we will see it as the following:

**Definition 1.2 (Quality assurance).** A planned and systematic pattern of all actions necessary to provide adequate confidence that an item or product conforms to established technical requirements [108].

Therefore, quality assurance includes all technique we use to build products in a way to increase our confidence as well as techniques to analyse products. This is expressed in the differentiation between *constructive quality assurance* and *analytical quality assurance*.

**Definition 1.3 (Constructive quality assurance).** All means to be used in constructing a product in a way so that it meets its quality requirements.

**Definition 1.4 (Analytical quality assurance).** All means of analysing the state of the quality of a product.

Hence, in analytical quality assurance, we want to find problems in the products. We colloquially call these problems in software a *bug*. The following definitions clarify the associated terms as we use them in this book. We start with problems visible to the user.

**Definition 1.5 (Failure).** A failure is an incorrect output of a software visible to the user.

A simple example is the output "12" of a calculator when the correct output is "10". The notion of a *failure* is the most intuitive term and it is easy to define (in most cases). During the run of a software something goes wrong and the user of the software notices that. Hence, output is anything that can be visible to the user, e.g. command line, GUI or actuators in an embedded system. Incorrect output can range from a wrong colour on the user interface over a wrong calculation to a total system crash. No output where output is expected is also considered to be incorrect.

The problem with defining a failure starts when we look in more detail at the term *incorrect*. It is difficult to define formally. Most of the time it is sufficient to say that a result is incorrect if it does not conform to the requirements as in the manufacturing approach to quality (see Sect. 1.3). Sometimes, however, the user did not specify suitably or this part of the software is underspecified. For our purposes, we follow the hybrid approach of requirements conformance and user expectation. For most cases it is sufficient to understand failures as deviations from the requirements specification. For requirements problems, however, it is necessary to include the user expectation.

**Definition 1.6 (Fault).** A fault is the cause of a potential failure inside the code or other artefacts. Synonym: bug.

Having the definition of a failure, the definition of a fault seems to be simple at first glance. The concept of a fault turns out to be difficult, however. It is often not possible to give a concrete location of the source of a failure. When the fault is an omission it is something non-existing and it is not always clear where the fix should be introduced. Another problem constitutes interface faults, i.e. the interfaces of two components do not interact correctly. It is possible to change either of the components to prevent the failure in the future. Frankl et al. discuss this in [68]. They state that the term *fault* "is not precise, and is difficult to make precise". They define a fault using the notion of a *failure region*. It consists of failure points – input values that cause a failure – that all do not cause a failure after a particular change. This allows us defining a fault precisely, but it does not give us a relation to programmer mistakes and therefore any classification of faults becomes difficult. Hence, we will use our own more ambiguous definition given above. Note that we consider wrong or missing parts in requirements or design specifications also as faults; i.e. this term is not restricted to code.

**Definition 1.7 (Defect).** Defects are the superset of faults and failures.

The notion of a *defect* is handy if it is not important whether we are considering a fault or failure in a certain context. This is because there is always some kind of relationship between the two and at a certain abstraction layer it is useful to have a common term for both.

**Definition 1.8 (Mistake).** A mistake is a human action that produces a fault.

The prime example is an incorrect action on the part of a developer including also omissions. This is sometimes called *error*. While discussing the notion of *fault*, we already saw that it might be interesting to have the relation to the actions of the developers – the mistakes they make. This is important for classification purposes as well as for helping to prevent those kinds of faults in the future. Hence, we have extended the 2-layer model of faults and failures to a 3-layer model where mistakes cause faults and faults cause failures as depicted in Fig. 1.6.

**Definition 1.9 (Error).** An error is that part of the system state that may cause a subsequent failure [4].

This definition extends the 3-layer model of mistakes, faults and failures to four layers with errors between faults and failures. Figure 1.6 gives an overview of the terms and the different layers. Intuitively, when running the program, a fault in the code is executed. Then the software does something not expected by the programmer and reaches some erroneous state. It has not yet produced a failure. At this stage so-called *fault-tolerance* mechanisms can take countermeasures to avoid the failure. Also the error might not lead to a failure because it has no consequence

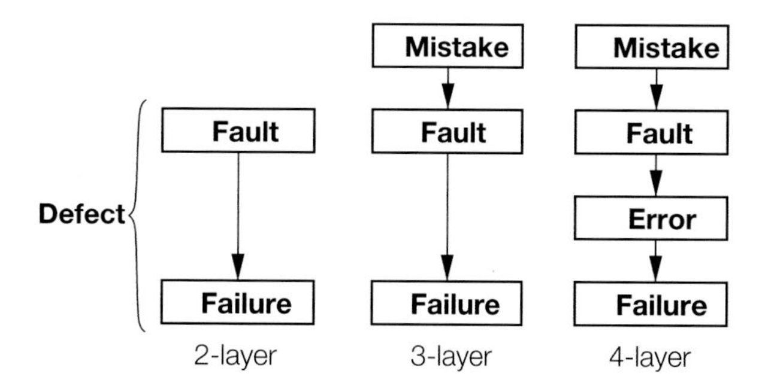


**Fig. 1.6** Overview of the terms related to defects [202]

on the user-visible behaviour, or it is masked by another error. In other cases, however, the erroneous state becomes visible to the user and hence results in a failure.

This differentiation in four layers of defects from the mistakes of developers over faults in artefacts, which lead to errors and finally to failures, allows us to clearly describe quality problems and especially discuss defect detection techniques. As we will discuss in Chap. 4, for example, testing can only reveal failures while inspections find faults. Depending on our goals, we can choose the model with the appropriate number of layers.

To find defects in our artefacts as part of analytical quality assurance, we apply verification and validation techniques (short V&V). These two terms are also often confused. In short, verification is concerned with whether we made the product right and validation if we made the right product.

**Definition 1.10 (Verification).** Verification is the activity to evaluate whether a given work result fits to its specified requirements.

**Definition 1.11 (Validation).** Validation is the activity to evaluate if a work result and its requirements fit to the stakeholder expectation.

As part of V&V, we differentiate between two different types of analyses to find defects: *static analysis* and *dynamic analysis*.

**Definition 1.12 (Static analysis).** The process of evaluating a system or component based on its form, structure, content or documentation [108].

Hence, static analysis means assessing the software without executing it. Examples are reviews, inspections or static analysis tools.

**Definition 1.13 (Dynamic analysis).** The process of evaluating a system or component based on its behaviour during execution [108].

Dynamic analysis is then the opposite of static analysis: executing the software to analyse it. Examples are testing or application scanners.

### 1.4.2 Quality Models

We have used the term *quality model* several times in this book already in connection with quality factors or quality characteristics. We will discuss quality models with a rather general meaning. You will read about the history and a richer discussion of the different purposes and application scenarios later in Sect. 2.1.

**Definition 1.14 (Quality model).** A model with the objective to describe, assess and/or predict quality.

Many quality models describe a decomposition of the general product quality into sub-qualities to make them better understandable and manageable.

**Definition 1.15 (Quality Factor).** A management-oriented attribute of software that contributes to its quality. [108] Synonyms: quality characteristic, quality aspect, quality attributes, qualities.

In the context of quality models, we are also confronted with the terms *quality requirements* and *quality goals* describing demands to the quality of a software system. We distinguish them by seeing quality goals as more abstract while quality requirements should be concrete and measurable.

**Definition 1.16 (Quality Goal).** An abstract demand onto a quality factor of a software product.

**Definition 1.17 (Quality Requirement).** A concrete and measurable demand on a specific product factor that has an impact onto a quality goal or quality factor of a software product.

### 1.4.3 Quality Evaluation

Evaluating a product for its current state of quality is an important part of quality control to derive corrective actions. It can make use of many different techniques with a focus on analytical quality assurance.

**Definition 1.18 (Quality evaluation).** Systematic examination of the extent to which an entity is capable of fulfilling specified requirements [108]. Synonym: quality assessment.

In many cases, the fulfilment of specified quality requirements will have to be quantitative. For example, if there is a performance requirement that a certain task has to be finished in 2 s, we have to measure the task completion. Therefore, in quality evaluation, we often deal with measures.

**Definition 1.19 (Measure).** Variable to which a value is assigned as the result of measurement [108]. Synonym (in software engineering): metric.

A measure represents some property of the product which we try to quantify using measurements.

**Definition 1.20 (Measurement).** Set of operations having the object of determining a value of a measure [108].

Measures and measurement will later be a central means to analyse and, finally, evaluate the software product. In our quality models, we also include the definition of measures.

### *1.4.4 Software Evolution*

Software product quality control is already interesting in the initial development of a new software product. Its full potential, however, is only realised during its maintenance. The term *software maintenance* is a bit odd as maintenance for mechanical systems usually refers to replacing worn out parts. This is not the case for software.

**Definition 1.21 (Software maintenance).** The process of modifying a software system or component after delivery to correct faults, improve performance or other attributes, or adapt to a changed environment [108].

Hence, software maintenance is the development after the initial release. Working on an existing, possibly long-lived, system is very challenging and, as we have seen above, software has a tendency to age. Hence, we will talk in various parts in this book about maintenance. If I want to refer to the complete life cycle, initial development and maintenance, I will use the term *evolution*.

**Definition 1.22 (Software evolution).** The whole life cycle of a product from conception and development over maintenance to retirement.

## 1.5 Overview of the SQuaRE Series of Standards

The international standard most directly applicable to software product quality control is the SQuaRE series of standards of the *International Organization for Standardization (ISO)*. SQuaRE stands for *Software product Quality, Requirements and Evaluation*. It is a long-running project at ISO to consolidate and replace the old ISO/IEC 9126 [107] which used to be the main standard for software product quality. The SQuaRE series increases the consistency with other ISO standards focused on measurement [94] and process quality [93]. In the end, the aim is to have one coherent set of standards to model, specify, measure and evaluate software product quality. At the time of writing of this book, the main standards for the planned divisions of the series have been approved. There are several more detailed standards yet missing, however. In particular, some standards describing measures to be used are still in the works. We will look at the major division of this series of standards and summarise their contents. I will later in the book refer to them where appropriate.

### 1.5.1 Quality Management

The first division of the series of standards is called *quality management*. The first standard in this division, ISO/IEC 25000 [95], which is also the oldest one of the series, gives an overview of the complete series and guidance on how to use them. It defines the scope of the series using a set of stakeholders including developers, acquirers and independent evaluators of software products. The main content are terms and definitions used in the other standards of the series. Several of them are already outdated, however, as they still refer to internal and external quality while this has been merged into product quality.

The second standard, ISO/IEC 25001 [96], relates more directly to the name of the division and describes requirements on the process for quality requirements and evaluation. It describes the relevant activities consistently with other ISO standards on the system life cycle.

### 1.5.2 Quality Model

The core of the whole series is the quality model in ISO/IEC 25010 [97]. As quality models for usage in quality control are also at the centre of this book, we will look at this standard in detail in Sect. 2.3. The standard describes a quality model framework, which is the *meta-model* of the quality models that should be used conforming to the standard. In summary, it defines that quality should be broken down in several layers of *quality characteristics* which is what we previously defined as *quality factors*. In addition, the standard defines two quality models for software: *product quality* and *quality in use*. For both of these models, it also contains definitions of the corresponding quality characteristics. Both models are slightly modified versions of the quality models from the older standard ISO/IEC 9126 [107].

A new quality model for data was defined in ISO/IEC 25012 [98]. It conforms to the quality model framework of ISO 25010 and defines the quality characteristics which are seen as important for data quality.

### 1.5.3 Quality Measurement

Probably the most complicated part of quality evaluation is measurement. Hence, there is a separate division for *quality measurement*. ISO/IEC 25020 [99] describes

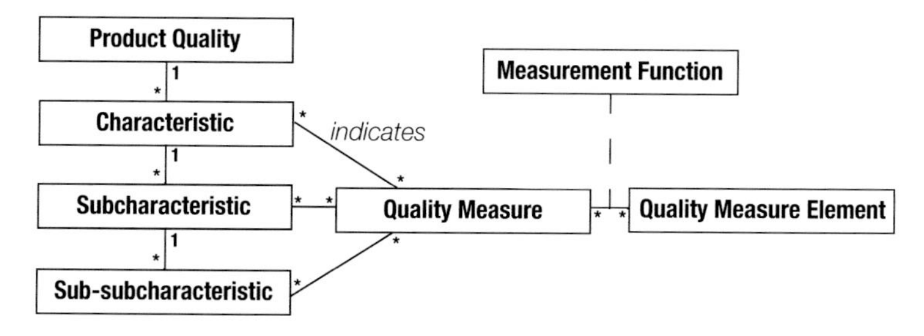


**Fig. 1.7** Software product quality measurement reference model of ISO/IEC 25020

what is measurement in the context of quality evaluation. Furthermore, it defines several basic terms such as *base measure*, a measure directly collectable from an artefact, and *derived measure*, a measure calculated from base measures. A central term is the *quality measure element*: It is a base measure or derived measure with which we construct higher-level *quality measures*. These quality measures are then indicators for quality characteristics or sub-characteristics. This results in the *software product quality measurement reference model* (SPQM-RM). My interpretation of this is shown in Fig. 1.7.

In addition, the standards define some requirements necessary to conform to the standard. For example, we have to select and document quality measures together with the criteria for selecting them. These criteria include the relevance to the information needs of the stakeholders or the ease of data collection. Overall, the standard remains unclear, however, why these several layers of quality measure elements and quality measures are necessary. We will later only discuss measures which can use other measures.

ISO/IEC 25021 [100] details the concept of *quality measure elements*. Each of the elements has an associated *measurement method* for the property of the target entity to be measured. The measurement method consists of the operations necessary to get the data for the measure element. The target entity is the "thing" of relevance to be measured. The standard gives a format how to document measure elements as well as example measure elements such as the number of accessible functions, the number of user problems or effort. Although this standard gives some very concrete quality measure elements, it relies on further standards and work specific for companies and projects to further add and detail to it.

## 1.5.4 *Quality Requirements*

One major idea of using quality models has always been to categorise quality requirements. This is also mentioned in the division and corresponding standard ISO/IEC 25030 [101]. It describes that we can formulate quality requirements categorised along the quality characteristics by using their quality measures.

To specify a quality requirement, we simply need to set a target value for the quality measure. It also includes a more general discussion on the need to define system boundaries and stakeholders in the process of specifying the quality requirements. It ends with requirements on quality requirements such as that the requirements should be uniquely identified and they should be verifiable. The standard captures concisely the main process for specifying quality requirements, however, without a clear guidance on how and when to apply it.

### *1.5.5 Quality Evaluation*

The final division of the series is concerned with *quality evaluation*. This is different from quality measurement which only collects data and applies measurement functions. The quality evaluation adds a level of interpretation to the data: To what extent does the product under evaluation meet its specified criteria? In particular, we usually transform the measurement into a *rating level* for easier understanding. For that, the standard ISO/IEC 25040 [102] defines a *software product quality evaluation reference model* which includes the evaluation process as well as roles such as supplier or independent evaluator. An important part of the standard are evaluation records which are meant for documenting evidence of the performed activities and the their results.

ISO/IEC 25041 [103] makes this evaluation process more concrete in an evaluation guide for developers, acquirers and independent evaluators. Finally, for some reason, *recoverability* has already its own evaluation module showing how to evaluate this quality characteristic in ISO/IEC 25045 [104].

## 1.6 Summary and Outline

In this chapter, we have motivated why product quality has become a crucial aspect of software today. We have discussed and defined product quality and related terms such as maintenance or defects and introduced the SQuaRE series of international standards for software product quality. Yet, what does this mean for you? How will that help you in implementing effective product quality control for your software systems? Although we aimed at making the vague concepts around quality more precise, this chapter probably leaves you with more questions than answers. The next chapters will give you these answers.

We have discussed the definition of quality and quality factors with the difficulty to break quality down into something concrete. Chapter 2 will introduce you to quality models and measures to achieve that, because they will be the basis for quality control. Using the quality models, we will discuss quality planning as part of the quality management process. Chapter 3 will guide you to specify quality requirements using quality models and to plan validation and verification.

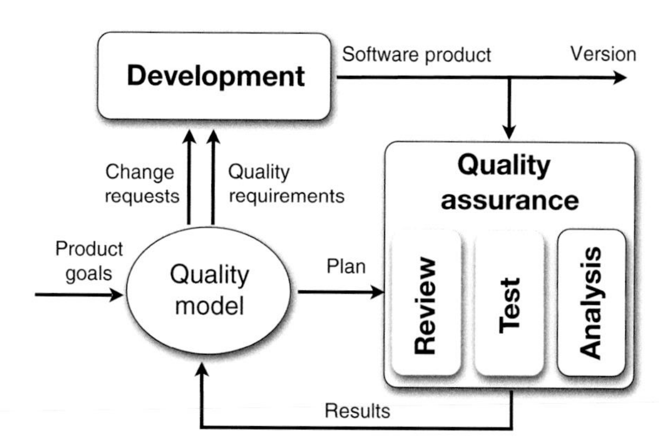


**Fig. 1.8** Quality control loop

After planning the product quality of our software system, we dive into the actual quality control. We do that in Chap. 4. We will first introduce the quality control loop as shown in Fig. 1.8. It shows how we continuously evaluate the quality of the software product in cycles for each new version. We will discuss various quality assurance techniques and how we can employ them in the quality control loop.

Finally, all of the issues and solutions we will show you are not only based on scientific rigour, but we also put them to practical test and conducted empirical studies. You will find concrete examples of several facets of quality control as we applied them with industrial partners in Chap. 5.

# Chapter 2
# Quality Models

In this chapter, after discussing existing quality models and putting them into context, I will introduce basics of software measures and details of the ISO/IEC 25010 quality model. The main part of this chapter constitutes the quality modelling approach developed in the research project Quamoco, how to maintain such quality models and three detailed examples of quality models.

## 2.1 Quality Models Set into Context

This whole chapter is dedicated to software quality models, because they constitute the foundation of software product quality control as we understand it in this book. There is a lot of existing literature and approaches for software quality models. Therefore, we start with setting our notion of quality models into context with a short history, definitions and classifications and a description of an example for each class of quality models.

### *2.1.1 A Short History of Software Quality Models*

Quality models have been a research topic for several decades and a large number of quality models have been proposed [125]. We cannot cover the area completely, but we will look into three historical threats in the development of quality models: hierarchical models, meta-model-based models and implicit quality models.

#### Hierarchical Quality Models

The first published quality models for software date back to the late 1970s, when Boehm et al. [24] as well as McCall, Richards and Walter [149] described quality

S. Wagner, *Software Product Quality Control*, DOI 10.1007/978-3-642-38571-1_2,

characteristics and their decomposition. The two approaches are similar and use a hierarchical decomposition of the concept *quality* into quality factors such as *maintainability* or *reliability*. Several variations of these models have appeared over time. One of the more popular ones is the FURPS model [77] which decomposes quality into functionality, usability, reliability, performance and supportability. Besides this hierarchical decomposition, the main idea of these models is that you decompose quality down to a level where you can measure and, thereby, evaluate quality.

This kind of quality models became the basis for the international standard ISO/IEC 9126 [107] in 1991. The standard defines a standard decomposition into quality characteristics and suggests a small number of metrics for measuring them. These metrics do not cover all aspects of quality, however. Hence, the standard does not completely operationalise quality. The successor of ISO/IEC 9126, the new standard ISO/IEC 25010 [97], changes a few classifications but keeps the general hierarchical decomposition.

In several proposals, researchers have used metrics to directly measure quality characteristics from or similar to ISO/IEC 9126. Franch and Carvallo [66] adapt the ISO quality model and assign metrics to measure them for selecting software packages. They stress that they need to be able to explicitly describe "relationships between quality entities". Van Zeist and Hendriks [219] also extend the ISO model and attach measures such as *average learning time*. Samoladas et al. [185] use several of the quality characteristics of ISO/IEC 9126 and extend and adapt them to open source software. They use the quality characteristics to aggregate measurements to an ordinal scale. All these approaches reveal that it is necessary to extend and adapt the ISO standard. They also show the difficulty in measuring abstract quality characteristics directly.

Ortega, Perez and Rojas [169] take a notably different view by building what they call a *systemic quality model* and using *product effectiveness* and *product efficiency* as the basic dimensions to structure similar quality characteristics as in the other approaches, such as reliability or maintainability.

Experimental research tools (e.g. [142, 187]) take first steps towards integrating a quality model and an assessment toolkit. A more comprehensive approach is taken by the research project Squale [155]. Here, an explicit quality model is developed that describes a hierarchical decomposition of the ISO/IEC 9126 quality characteristics. The model contains formulas for aggregating and normalising metric values. Based on the quality model, Squale provides tool support for evaluating software products. The measurements and the quality model are fixed within the tools.

Various critiques (e.g. [2,51]) point out that the decomposition principles used for quality characteristics are often ambiguous. Furthermore, the resulting quality characteristics are mostly not specific enough to be measurable directly. Although the recently published successor ISO/IEC 25010 has several improvements, including a measurement reference model in ISO/IEC 25020, the overall critique is still valid because detailed measures are yet missing. Moreover, a survey done by us [213] shows that less than 28 % of the companies use these standard models and 71 % of them have developed their own variants. Hence, there is a need for customisation.

### Meta-Model-Based Quality Models

Starting in the 1990s, researchers have been proposing more elaborate ways of decomposing quality characteristics and thereby have built richer quality models based on more or less explicit meta-models. A meta-model, similar as in the UML,[1] is a model of the quality model, i.e. it describes how valid quality models are structured. In addition, the proposals also differ to what extent they include measurements and evaluations.

An early attempt to make a clear connection between measurements and the quality factors described in the hierarchical quality models was made in the ESPRIT project *REQUEST*. They developed what they called a *constructive quality model* COQUAMO [120, 124]. They see the quality factor as central in a quality model and argue that each factor should be evaluated differently throughout different development phases and, therefore, have different metrics. They also differentiate between application-specific and general qualities. For example, they see reliability as important for any software system, but security is application specific. COQUAMO aimed strongly at establishing quantitative relationships between quality drivers measured by metrics and factors. We will discuss COQUAMO in more detail below as an example for multi-purpose models.

With some similarity is the also similarly named COQUALMO [37]. It uses drivers, similar to the effort model COCOMO [23], and describes a defect flow model. In each phase, defects are introduced by building and changing artefacts and removed by quality assurance techniques. For example, requirements defects are introduced because of misunderstood expectations of the stakeholders. Some of these requirements defects are then removed by reviews or later testing. An estimation of the drivers enables COQUALMO to estimate the defect output.

Without an explicit meta-model but with a similar focus on establishing statistical relationships, there is a plethora of various very specific quality models. They use a set of measure which are expected to influence a specific quality factor. For example, the *maintainability index* (MI) [41] was a set of static code measures combined so that the resulting index should give an indication of the maintainability of the system. Another example is reliability growth models [139] which relate test data with the reliability of the software.

Dromey [54] published a quality model with a comparably elaborate meta-model in which he distinguishes between *product components*, which exhibit *quality-carrying properties*, and externally visible *quality attributes*. Product components are parts of the product, in particular the source code of the software. For example, a variable is a product component with quality-carrying properties such as *precise*,

[1] http://www.uml.org.

*assigned* or *documented*. These are then set into relation to the quality factors of ISO/IEC 9126.

Kitchenham et al. [21, 121] built the SQUID approach on the experiences from COQUAMO. They acknowledge the need for an explicit meta-model to describe the increasingly complex structures of quality models. Although they state in [121] that the REQUEST project concluded that "there were no software product metrics that were, in general, likely to be good predictors of final product qualities" and that there is still no evidence on that, it is still useful to model and analyse influences to quality. Hence, they propose to monitor and control "internal" measures which may influence the "external" quality. Their quality meta-model includes *measurable properties* that can be *internal software properties* or *quality subcharacteristics*. The internal software properties influence the quality subcharacteristics and both can be measured. They also introduce the notion to define a target value for the measurable properties for a quality requirements specification.

Bansiya and Davis [7] built on Dromey's model and proposed QMOOD, a quality model for object-oriented designs. They described several metrics for the design of components to measure what they call *design properties*. These properties have an influence on quality attributes. They also explicitly mentioned tool support and describe empirical validations of their model.

Bakota et al. [6] emphasised the probabilistic nature of their quality model and quality assessments. They introduced *virtual quality attributes* which are similar to the internal software properties of SQUID. The quality model uses only nine low-level measures which are evaluated and aggregated to probability distributions. Our experience has been, however, that practitioners have difficulties interpreting such distributions.

All these meta-model-based quality models show that the complex concept of *quality* needs more structure in quality models than abstract quality characteristics and metrics. They have not established a general base quality model, however, which you can just download and apply.

### Statistical and Implicit Quality Models

For various quality factors, statistical models have been proposed that capture properties of the product, process, or organisation and estimate or predict these quality factors. A prominent example of such models are *reliability growth models* [139, 158]. They transfer the idea of hardware reliability models to software. The idea is to observe the failure behaviour of a software, for example, during system testing, and predict how this behaviour will change over time. Similar models are the *maintainability index* (MI) [41], a regression model from code metrics or Vulture [166], a machine learning model predicting vulnerable components based on vulnerability databases and version archives.

Although they usually do not explicitly say so, many quality analysis tools use some kind of quality model. For example, tools for bug pattern identification (e.g. FindBugs, Gendarme or PC-Lint) classify their rules into different categories based on the type of problems they detect (dodgy code), what quality factor it might influence (e.g. performance) or how severe the problem is. Therefore, implicitly, these tools already define a lot what we would expect from a quality model: quality factors, measurements and influences. Usually, it is not made explicit, however, which makes a comprehensive quality evaluation difficult.

Dashboards can use the measurement data of these tools as input (e.g. QALab, Sonar or XRadar). Their goal is to present an overview of the quality data of a software system. Nevertheless, they often also lack an explicit connection between the metrics used and the required quality factors. Hence, explanations of the impacts of defects on software quality and rationales for the used metrics are missing.

Finally, also checklists used in development or reviews are a kind of quality model. They are usually not directly integrated with a quality model at all. They define properties of artefacts, most often source code, which have an influence on some quality factor. Sometimes the checklists make these influences explicit. Most often, they do not. This leaves the software engineers without a clear rationale why to follow the guidelines. Hence, explicit quality models can also help us to improve checklists and guidelines.

### 2.1.2 *Definitions and Classifications*

Software quality models are now a well-accepted means to support quality control of software systems. As we have seen, over the last three decades a multitude of very diverse models commonly termed "quality models" have been proposed. On first sight, such models appear to have little relation to each other although all of them deal with software quality. We claim that this difference is caused by the different purposes the models pursue [49]: The ISO/IEC 25010 is mainly used to *define* quality, metric-based approaches are used to *assess* the quality of a given system and reliability growth models are used to *predict* quality. To avoid comparing apples with oranges, we propose to use these different purposes, namely, *definition, assessment* and *prediction* of quality, to classify quality models. Consequently, we term the ISO/IEC 25010 as *definition model*, the maintainability index as *assessment model* and reliability growth models as *prediction models*.

Although *definition, assessment* and *prediction* of quality are different purposes, they are not independent of each other: It is hard to assess quality without knowing what it actually constitutes and equally hard to predict quality without knowing how to assess it. This relation between quality models is illustrated by the DAP classification shown in Fig. 2.1.

The DAP classification views prediction models as the most advanced form of quality models as they can also be used for the definition of quality and for its

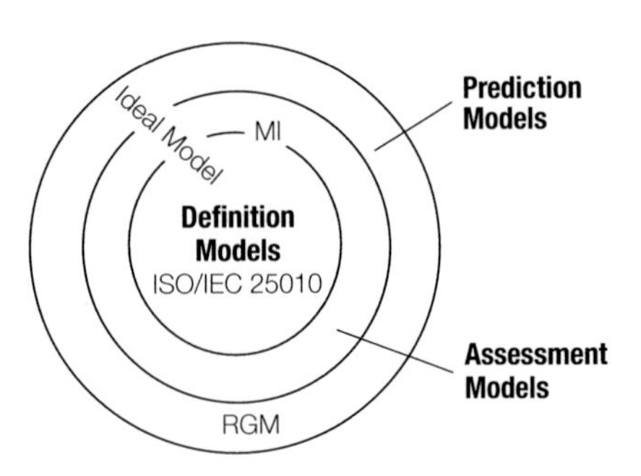


**Fig. 2.1** DAP classification for quality models [49]

assessment. This view only applies for ideal models, however. As Fig. 2.1 shows, existing quality models do not necessarily cover all aspects equally well. The ISO/IEC 25010, for example, defines quality but gives no hints for assessing it; the MI defines an assessment whose relation to a definition of quality is unclear. Similarly, reliability growth models perform predictions based on data that is not explicitly linked to an overall definition of quality.

To reflect the importance of the purpose, we propose a definition of quality models [49] that explicitly takes the purpose of the quality model into account. Due to the diversity of different models, we do not restrict the type of model to a specific modelling technique or formalism.

**Definition 2.1 (Quality Model).** A model with the objective to describe, assess and/or predict quality.

Independent of the modelling technique used to build a quality model, we consider the existence of a defined meta-model as crucial. Even though, in many cases, quality meta-models are not explicitly defined for existing quality models. A meta-model is required to precisely define the model elements and their interactions. It not only defines legal model instances but also explains how models are to be interpreted. Accordingly, we define [49]:

**Definition 2.2 (Quality Meta-Model).** A model of the constructs and rules needed to build specific quality models.

Finally, it must be defined how the quality model can be used in the development and evolution of a software system. This typically concerns the process by which a quality model is created and maintained and the tools employed for its operationalisation. We call this a *quality modelling framework*.

**Definition 2.3 (Quality Modelling Framework).** A framework to define, evaluate and improve quality. This usually includes a quality meta-model as well as a methodology that describes how to instantiate the meta-model and use the model instances for defining, assessing, predicting and improving quality.

Most of the remainder of this book will deal with the Quamoco[2] quality modelling framework. To give you a broader view on the existing quality models, we will first look, however, at some other examples for models for the different purposes.

### 2.1.3 Definition Models

*Definition* models are used in various phases of a software development process. During requirements engineering, they define quality factors and requirements for planned software systems [122] and thus constitute a method to agree with the customer what quality means [122]. During implementation, quality models serve as basis of modelling and coding standards or guidelines [54]. They provide direct recommendations on system implementation and, thus, constitute constructive approaches to achieve high software quality. Furthermore, quality defects that are found during quality assurance are classified using the quality model [54]. Apart from their use during software development, definitional quality models can be used to communicate software quality knowledge during developer training or student education.

As we will look into the ISO/IEC 25010 quality model in more detail in Sect. 2.3, we use the FURPS model [77] as an example here. FURPS stands for functionality, usability, reliability, performance, and supportability. It is a decomposition of software quality. Each of these quality factors is also further decomposed:

- Functionality
  - Feature set
  - Capabilities
  - Generality
  - Security
- Usability
  - Human factors
  - Aesthetics
  - Consistency
  - Documentation
- Reliability
  - Frequency/severity of failure
  - Recoverability
  - Predictability
  - Accuracy
  - Mean time to failure

[2] http://www.quamoco.de/.

- Performance
  - Speed
  - Efficiency
  - Resource consumption
  - Throughput
  - Response time
- Supportability
  - Testability
  - Extensibility
  - Adaptability
  - Maintainability
  - Compatibility
  - Configurability
  - Serviceability
  - Installability
  - Localisability
  - Portability

Hence, FURPS is a hierarchical definition model. The first four quality factors (FURP) are more aimed at the user and operator of the software, while the last quality factor (S) is more targeted at the developers, testers and maintainers. FURPS gives an alternative decomposition to the standard ISO/IEC 25010 which we will discuss in detail in Sect. 2.3. The main aim of FURPS is a decomposition and checklist for quality requirements. A software engineer can go through this list of quality factors and check with the stakeholders to define corresponding qualities. Therefore, it defines quality as basis for requirements. In addition, Grady and Caswell [77] describe various metrics that can be related to the quality factors for evaluating them. The main purpose of FURPS, however, is to *define* quality.

### 2.1.4 Assessment Models

*Assessment* models often extend quality definition models to evaluate the defined qualities of the definition model. During requirements engineering, assessment models can be used to objectively specify and control stated quality requirements [122]. During implementation, the quality model can be the basis for all quality measurements, i.e. for measuring the product, activities and the environment [54, 197, 202]. This includes deriving guidelines for manual reviews [51] and systematically developing and using static analysis tools [54, 174]. Thereby, we monitor and control internal measures that might influence external properties [122]. Apart from their use during software development, assessment models furthermore constitute the touchstone for quality certifications.

The *EMISQ* [174,176] method constitutes an assessment model based on the ISO standard 14598 for product evaluation [92]. It defines an approach for assessing "internal" quality attributes like maintainability and explicitly takes into account the expertise of a human assessor. The method can be used as is with a reference model that is a slight variation of the ISO/IEC 9126 model or customisations thereof. Consequently, the method's quality definition model is very similar to ISO/IEC 9126. It defines quality characteristics and exactly one level of subcharacteristics that are mapped to quality metrics, whereas one subcharacteristic can be mapped to multiple metrics, and vice versa. It uses as metrics results from well-known quality assessment tools like *PC-Lint*[3] and *PMD*.[4] Hence, they include not only classic numeric metrics but also metrics that detect certain coding anomalies. A notable property of the EMISQ method is that its reference model includes about 1,500 different metrics that are mapped to the respective quality characteristics. The approach also provides tool support for building customised quality models and for supporting the assessments.

### 2.1.5 Prediction Models

In the context of software quality, predictive models have (among others) been applied to predict the number of defects of a system or specific modules, mean times between failures, repair times and maintenance efforts. The predictions are usually based on source code metrics or past defect detection data.

A prediction of reliability is provided by *reliability growth models (RGMs)* that employ defect detection data from test and operation phases to predict the future reliability of software systems [139, 158]. Such models assume that the number of defects found per time unit decreases over time, i.e. the software becomes more stable. The idea then is that if we measure the failure times during system tests with an execution similar to future operation, we will be able to interpolate from them to the failure behaviour in the field. Figure 2.2 illustrates this example data. It shows the calendar time on the x-axis and the cumulated number of failures on the y-axis. Each occurred failure is shown as a cross in the diagram. The curved line is then a fitted statistical model of the failure data. This model goes beyond the already occurred failures and is, hence, able to predict the probable future occurrence of failure. This can then be expressed as reliability.

There are various difficulties in applying reliability growth models. First of all, we need to decide on a suitable statistical model that adequately represents the actual failure distribution. There are many different proposal and it is hard to decide beforehand which one is the best fit. In addition, time measurement is a problem because software does not fail just because clock time passes, but it has to be used.

---

[3]http://www.gimpel.com.

[4]http://pmd.sourceforge.net.

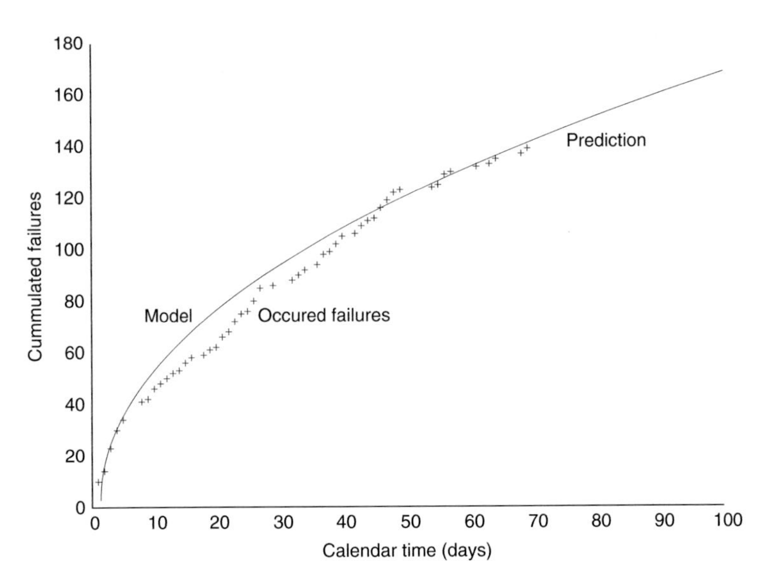


**Fig. 2.2** An example prediction of a reliability growth models

Hence, we need to measure appropriate time counts. Furthermore, a direct relation to a definition model is missing. Hence, if the reliability of our system is predicted as too low, it is not clear what we have to do to improve it. We will see a detailed example in Sect. 2.6.3.

### 2.1.6 *Multi-Purpose Models*

An ideal model in the DAP classification is a quality model that integrates all three purposes. Such a *multi-purpose model* has the advantage that we assess and predict on the same model that we use to define the quality requirements. It ensures a high consistency which might be endangered by separate definition and assessment models. One of the rare examples of a multi-purpose model is the already-mentioned *COQUAMO*.

COQUAMO was a big step forward in the 1980s to describe quality measures quantitatively and to relate measures of the product, process or organisation to measures of the product's quality using statistical models. It drew inspiration from COCOMO [23], Boehm's effort estimation model. Especially the prediction part of COQUAMO reflects this inspiration. Figure 2.3 shows the three main measures used in COQUAMO: *quality factor metrics*, *measures of quality drivers* and *quality indicators*. Each of these measures has its own purpose. The quality factor metrics measure quality factors directly. For example, the quality factor *reliability* could be measured by the *mean time to failure*. The quality factors with their quality

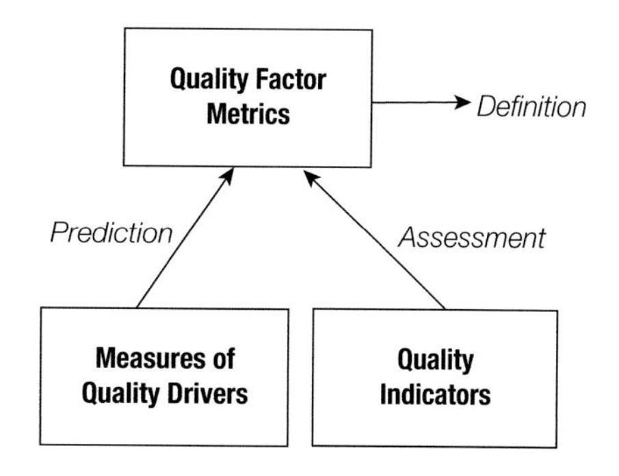


**Fig. 2.3** Definition, assessment and prediction in COQUAMO

factor metrics define quality and, together with target values, can be used to specify quality requirements. The quality factors are similar as in other, hierarchical quality models: usability, testability or maintainability.

Similar to COCOMO, we then can estimate quality drivers using *measures of quality drivers* for an early prediction of the final quality of the software. These quality drivers include product attributes, such as the quality requirements; process attributes, such as the process maturity; or attributes of the personnel, such as experience. COQUAMO establishes relationships between these quality drivers and the quality factors. The idea is that a project manager can predict and adjust the drivers early in a project so that the quality requirements are achievable.

The *quality indicators* are used for assessing the quality of the product during its development to monitor and control quality. These quality indicators measure attributes of the product directly, and through established statistical relationships to quality factors, they should give an indication of the current quality. For example, a quality indicator can be the number of called modules in a module or McCabe's cyclomatic complexity [147].

A further interesting concept lies in the different measurements of quality indicators for a quality factor over the life cycle of a software. In the beginning, we should use requirements checklists, which includes standards and metrics, then design checklists, coding checklists and, finally, testing checklists. This reflects that different artefacts produced over time can contain information we can use to analyse quality.

Overall, the difficulty in COQUAMO lies in the unclear relationships of the different types of measurements. There are no functional relationships from quality indicators, such as the number of called modules, and quality factor metrics, such as the mean time to failure. Only statistical relationships can be investigated and even they are very hard to establish as many contextual factors play a role. Later, Kitchenham et al. [121] concluded from COQUAMO that "there were no software product metrics that were, in general, likely to be good predictors of final product qualities". Nevertheless, there are influences from quality indicators and quality drivers to quality factors, but a direct prediction remains difficult.

### 2.1.7 Critique

All the existing quality models have their strengths and weaknesses. Especially the latter have been discussed in various publications. We summarise and categorise these points of criticism. Please note that different quality models are designed for different intentions and therefore not all points are applicable to all models. However, every quality model has at least one of the following problems [49].

#### General

One of the main shortcomings of many existing quality models is that they do not conform to an explicit meta-model. Hence the semantics of the model elements is not precisely defined and the interpretation is left to the reader.

Quality models should act as a central repository of information regarding quality, and therefore, the different tasks of quality engineering should rely on the same quality model. Today, most quality models are not integrated into the various tasks connected to quality. For example, the specification of quality requirements and the evaluation of software quality are usually not based on the same models.

Another problem is that today quality models do not address different views on quality. In the field of software engineering, the value-based view is typically considered of high importance [201] but is largely missing in current quality models [122].

Moreover, the variety in software systems is extremely large, ranging from huge business information systems to tiny embedded controllers. These differences must be accounted for in quality models by defined means of customisation. In current quality models, this is not considered [72, 119, 156].

#### Definition Models

Existing quality models lack clearly defined decomposition criteria that determine how the complex concept "quality" should be decomposed. Most definition models depend on a taxonomic, hierarchical decomposition of quality factors. This decomposition does not follow defined guidelines and can be arbitrary [27, 51, 121, 122]. Hence, it is difficult to further refine commonly known quality attributes such as availability. Furthermore, in large quality models, unclear decomposition makes locating elements difficult, since developers might have to search large parts of the model to assert that an element is not already contained in it. This can lead to redundancy due to multiple additions of the same or similar elements.

The ambiguous decomposition in many quality models is also the cause of overlaps between different quality factors. Furthermore, these overlaps are often not explicitly considered. For example, considering a denial of service attack, security is strongly influenced by availability which is also a part of reliability; code quality is an important factor for maintainability but is also seen as an indicator for security [8].

Most quality model frameworks do not provide ways for using the quality models for constructive quality assurance. For example, it is left unclear how the quality models should be communicated to project participants. A common method of communicating such information is guidelines. In practice, guidelines, which are meant to communicate the knowledge of a quality model, are not actually used. This is often related to the quality models itself; e.g. the guidelines are often not sufficiently concrete and detailed or the document structure is random and confusing. Also rationales are often not given for the rules the guidelines impose.

### Assessment Models

The already-mentioned unclear decomposition of quality factors is in particular a problem for analytical quality assurance. The given quality factors are mostly too abstract to be straightforwardly checkable in a concrete software product [27, 51]. Because the existing quality models neither define checkable factors nor refinement methods to get checkable factors, they are hard to use in measurement [69, 122].

In the field of software quality, a great number of measures have been proposed, but these measures face problems that also arise from the lack of structure in quality models. One problem is that despite defining measures, the quality models fail to give a detailed account of the impact that specific measures have on software quality [122]. Due to the lack of a clear semantics, the aggregation of measure values along the hierarchical levels is problematic. Another problem is that the provided measures have no clear motivation and validation. Moreover, many existing approaches do not respect the most fundamental rules of measurement theory (see Sect. 2.2) and, hence, are prone to generate dubious results [61].

It has to be noted that measurement is vital for any quality control. Therefore the measurement of the most important quality factors is essential for a effective quality assurance processes and for a successful requirements engineering.

### Prediction Models

Predictive quality models often lack an underlying definition of the concepts they are based on. Most of them rely on regression or data mining using a set of software metrics. Especially regression models tend to result in equations that are hard to interpret [62]. Also the results of data mining are not always easy to interpret for practitioners although some data mining techniques especially support this by showing decision trees.

Furthermore, prediction models tend to be strongly context dependent, also complicating their broad application in practice. Some researchers now particularly investigate local predictions [151]. It seems it is not possible to have one single set of metrics to predict defects, as found, for example, by Nagappan, Ball and

Zeller [163]. Many factors influence the common prediction goals and especially which factors are the most important ones varies strongly. Often these context conditions are not made explicit in prediction models.

#### Multi-Purpose Models

Although multi-purpose models provide consistency between definition, assessment and prediction, they are not yet widely established. The REQUEST project, which developed COQUAMO, took place in the 1980s. Despite that, thirty years later there is no established integrated multi-purpose quality model for software. A problem encountered in the REQUEST project was that it was difficult to find clear relationships between metrics as in other prediction models. Furthermore, I suspect that there is a large effort associated with building, calibrating and maintaining such a model. Finally, it might have also played a role that information about these old models is scarce and not easily available to software engineers.

### *2.1.8 Usage Scenarios*

In summary and to understand how we want to use software quality models, we will briefly discuss the main usage scenarios [49]:

*Definition* models are used in various phases of a software development process. During requirements engineering, they define quality factors and requirements for planned software systems [122, 204] and, thus, constitute a method to agree with the customer what quality means [122]. During implementation, quality models serve as basis of modelling and coding standards or guidelines [54]. They provide direct recommendations on system implementation and, thus, constitute constructive approaches to achieve high software quality. Furthermore, quality defects that are found during quality assurance can be classified using the quality model [54]. Apart from their use during software development, quality definition models can be used to communicate software quality knowledge during developer training or student education.

*Assessment* models often naturally extend quality definition model usage scenarios to control compliance. During requirements engineering, assessment models can be used to objectively specify and control stated quality requirements [122]. During implementation, the quality model is the basis for all quality measurements, i.e. for measuring the product, activities and the environment [54, 197, 202]. This includes the derivation of guidelines for manual reviews [51] and the systematic development and usage of static analysis tools [54, 174]. During quality audits, assessment models serve as a basis of the performed audit procedure. Thereby, internal measures that might influence external properties are monitored and controlled [122]. Apart from their use during software development, assessment models furthermore constitute the touchstone for quality certifications.

*Prediction* models are used during project management. More specifically, such models are used for release planning and in order to provide answers to the classical "when to stop testing" problem [157]. *Multi-purpose* models, finally, combine all the above usage scenarios.

### 2.1.9 Summary

An impressive development of quality models has taken place over the last decades. These efforts have resulted in many achievements in research and practice. As an example, consider the field of software reliability engineering that performed a wide as well as deep investigation of reliability growth models. In some contexts these models are applied successfully in practice. The developments in quality definition models led to the standardisation in ISO/IEC 25010 that defines well-known quality factors and will serve as the basis for many quality management approaches. It even integrates with a quality evaluation method in ISO/IEC 25040, so that the international standard describes a multi-purpose quality model to some degree.

The whole field of software quality models, however, remains diverse and fuzzy. There are large differences between many models that are called "quality models". Moreover, despite the achievements made, there are still open problems, especially in the adoption in practice. Because of this, current quality models are subject to a variety of points of criticism that will keep further generations of software quality researchers busy.

## 2.2 Software Measures

The area of software metrics has been under research from the early days of software engineering. It describes how to measure this abstract thing "software", hopefully in a way that helps the software engineers to develop software better and faster. As we talk about measurement, we adopted the term "measure" instead of "metric" in most parts of the book, but we will use it interchangeably. In this section, we discuss the basics of software metrics with interesting properties and scales and then focus on aggregation because that is very important for quality evaluation.

### 2.2.1 Basics

Measurement is the mapping from the empirical world to the formal world. In the empirical world, there are entities (things) that have certain attributes. These attributes can be expressed with measures from the formal world. Measurement

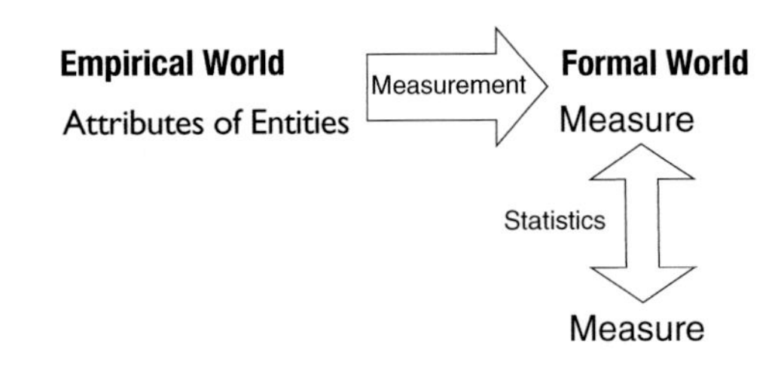


**Fig. 2.4** The general concepts of measurement and statistics

theory is therefore responsible for arguing about the relationship between reality and measures. For example, the table in my office is certainly an entity of reality and has the attribute *height*. Measurement maps this real attribute of the table to the formal world by stating that the height is 71 cm. Transferred to software this means that the entity *source code* has the attribute *length* which we can measure in lines of code (LOC). This relationship is depicted in Fig. 2.4.

Why would we be interested in this transformation into the formal world? Because we now can analyse the measures and, in particular, the relationships between different measures. Describing the relationships between those different measures is the task of statistical theory. For example, we can investigate and describe the relationship between the height of tables and their users. Grown-up "table users" will have, on average, higher tables than children in their rooms or in the class room. Similarly, software with more LOC has probably taken more effort (e.g. measured in person month) than software with less LOC.

We will come across some statistical theory below in the context of aggregation. Also for measurement, there is a well-proven measurement theory that helps us in avoiding mistakes in measuring and interpreting measurements. We will in particular discuss scales and important properties of measures. Scales are probably the most important part of measurement theory, because they can help us to avoid misinterpretations.

### Scales

In principle, there is an unlimited number of scales possible for software engineering measures. It usually suffices, however, to understand five basic scales to be able to interpret most measures:

1. Data that only gives names to entities has a *nominal scale*. Examples are defect types.
2. If we can put the data in a specific order, it has an *ordinal scale*. Examples are ratings (high, medium, low).
3. If the interval between the data points in that order is not arbitrary, the scale is *interval*. An example is temperature in Celsius.
4. If there is a real 0 in the scale, it is a *ratio* scale. An example are LOC.

5. If the mapping from the empirical world is unique, i.e. there is no alternative transformation, it is an *absolute* scale. An example is the ASCII characters in a file.

To understand in which of these scales your measures are is necessary to do the right interpretation. For example, temperature measured in Celsius is in an interval scale. The intervals between degrees Celsius are not arbitrary but clearly defined. Nevertheless, it does not make sense to say "It is twice as hot as it was yesterday!" if it has 30 degrees today and it had 15 degrees yesterday. Another perfectly valid unit for temperature is Fahrenheit in which, for the same example, it would be 59 Fahrenheit yesterday and 86 Fahrenheit today. That is not "twice as hot". The reason is that the 0 in these scales was artificially chosen. In contrast, LOC has a real 0 when there are no lines of code. Therefore, it is permissible to say "This software is twice as large as the other!"

Therefore, the scales define what is permissible to do with the data. We will use the scales later to discuss what aggregations are possible below in detail. For example, for measures with a nominal scale, we cannot do much statistical analysis but count the numbers of the same values or find the most often occurring value (called the *mode*). For ordinal data, I have an order and, therefore, I can find an average value which is the value "in the middle" (called the *median*). From an interval scale on, we can calculate a more common average, the *mean*, summing all the values and dividing by the number of values. With ratio and absolute scales, we can use any mathematical and statistical techniques available.

### Properties of Measures

Apart from the scale of a measure, we should consider further properties to understand the usefulness and trustworthiness of a measure. Very important desired properties of measures are reliability and validity. *Reliability* means in this context that the measure gives (almost) the same result every time it is measured. *Validity* means that its value corresponds correctly to the attribute of the empirical entity. Different possibilities for the reliability and validity of a measure are illustrated in Fig. 2.5. A measure is neither reliable nor valid if it produces a different value every time it is measured and none of these describes the attribute of the empirical entity well (left in Fig. 2.5). A measure can be reliable by producing similar results in every measurement but not be valid because the reliable measurements do not correspond to the empirical reality (middle). Finally, we want a measure which is reliable and valid (right). These two properties are expected to be achieved with any measure we define. Validity is often difficult to show and reliability can be tricky for manually collected measures which are often subjective.

In addition, there are further properties of measures which are also desired but not always possible to be achieved. As already discussed above, the reliability of a measurement can be problematic for subjective measures. Therefore, we aim for *objectivity* in measures meaning that there is no subjective influence in

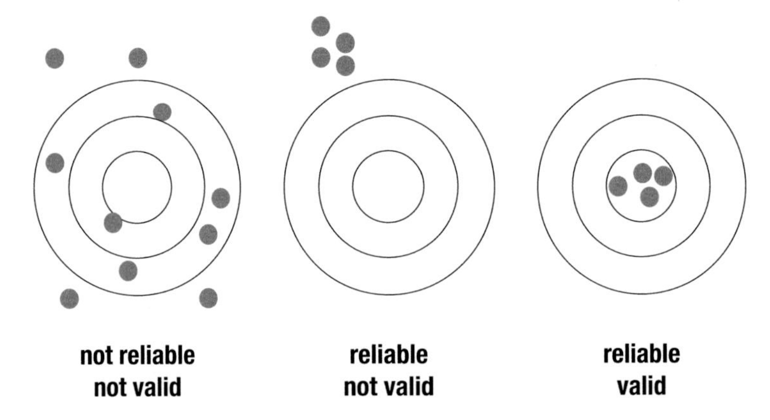


**Fig. 2.5** Reliability and validity of measures

measurement. Next, we want to be able to use the measure in the formal world and compare it to other measures (*comparability*). This requires a suitable scale for the measure (*standardisation*). Moreover, we can measure countless things in a software development project, but the *usefulness* of those measure should be ensured in the sense that they fulfil practical needs. Finally, *economy* is also a desired property for measures. It is helpful to be able to collect them with low cost. Then, we can collect and analyse often and early. We have observed, however, that in many cases the more useful the measures are, the less economic they are.

## 2.2.2 Aggregation

A plethora of measures for software has been proposed that use numerous properties and aim at various purposes. Several of these measures are independent of the granularity of the measured software entity. For example, lines of code can be measured for a whole system or for subcomponents only. Many of these measure, however, are "local", i.e. measure properties that are only useful for a specific part of the software. For example, the depth in the inheritance tree [34] can only be calculated for a single class in an object-oriented system. Aggregating these measures is not straightforward but necessary for quality evaluation. We will introduce here how we can apply aggregation operators to software measures and provide a collection of possible operators for your aggregation tasks. These operators are categorised according to the suitability for different purposes. We also describe the implementation of aggregation in a dashboard toolkit and use several of the operators in an example.

To have a model or goal for a measure is common in research to ensure its usefulness [10, 20, 57]. Building these models often includes the aggregation of measures to define higher-level measures. For example, it can be interesting to aggregate measures defined for classes to the system level to be able to compare different systems. As discussed in [86], the measurement system "Emerald aggregates the metrics [...] to a level specified by the user. The aggregation levels include source file, module, test configuration, development group, release, and major subsystem".

This point is also emphasised by Blin and Tsoukiàs [20]: "...the definition of an evaluation model implies a number of choices including the aggregation procedures to be used. Such choices are not neutral and have to be rigorously justified. Otherwise, results are very difficult to understand and to exploit, when they are not totally surprising". Hence, aggregation is an important part of any measurement system and a reoccurring task in any measurement method.

However, as Blin and Tsoukiàs also state in [20]: "It is surprising how often the choice of the aggregation operator is done without any critical consideration about its properties". For example, in the IEEE standard 1061 [87], only weighted sums are mentioned for aggregation. Yet, the choice of the right aggregation operator has a strong influence on the results and hence needs to be justified carefully.

### Aggregation Purposes

A justification of the suitability of a specific aggregation operator is only possible on the basis of the purpose of the aggregation. It is completely different if the current state of a system should be assessed or whether its hot spots should be identified. A complete list of aims of aggregation (and thereby measurement) is difficult to work out, but we identified five different aims that cover most cases. This categorisation is partly similar to the categorisation of purposes of quality models in Sect. 2.1. Note that they also build on each other and that more sophisticated analyses presuppose the simpler ones.

#### Assessment

The first and most common purpose of collecting data about a system is to assess its current state. This starts with simple measures such as size or age but comes quickly in extremely complex areas such as the reliability or the maintainability of the system. Important in assessment is that we do not take the future into account, at least not directly. Nevertheless, when determining a size trend in the assessment, it might have implications for the future as well. We are not interested in how the size grew over the time the system has existed, however. This is what in software reliability engineering is often called *estimation* [139].

There are three different modes in which assessments of software can be done. First, a singular, unique "health check" is a possibility in which the project or quality manager wants an overview of the system and of a specific set of its properties. Second, in process models for software development, there are often defined *quality gates* (Sect. 4.1) where the quality of certain artefacts needs to be checked. Third, it is possible to implement suitable assessments as part of the daily or weekly build of the software. This way, problems are detected early and can be confronted when they are still relatively cheap. For example, in our continuous quality control, several measures are routinely measured to check the current quality.

### Prediction

When assessments are in place and possible for certain software properties, the next desired step is to predict the future development of these properties. This is what most software reliability models are concerned with [139]. Here, we use information about the current state of the system, possibly enriched with historical data about the development of measures, to determine a probable future state of the system. Hence, usually there are probabilistic models involved in this prediction process. In prediction, we might need to aggregate the predicted results from subsystems to systems, for example.

As mentioned, reliability growth models are well-known examples that use historical defect data to predict the reliability in the field [136, 159, 207]. Another example in the context of maintenance is models that predict the future maintenance efforts for a software system [81, 221].

### Hot Spot Identification

Often, it is interesting not only to measure a certain property now or predict its development but also to detect specific areas of interest, denoted as "hot spots" here. This is especially important after the assessment and/or prediction to derive countermeasures against the detected problems. For a specific instruction for improvement, we need to know where to start with the improvement. Usually, we want to know where is the biggest problem and hence the largest return on investment when there is improvement.

For example, there are various methods for identifying fault-prone components [162, 208] using different approaches based on software measures. Also here the problem of aggregation can easily be seen: Is a component fault prone when its subcomponents are fault prone?

### Comparison

Another important purpose is to compare different components or systems based on certain (aggregated) measures. In principle, comparisons are simply assessments of more than one software. The difficulty lies in comparability of the measures used for comparison, however. Not all measures are independently useful without having its context. Hence, normalisations need to take place for this. A common example is normalisations by size as in the defect density that measures the average defects per LOC.

Comparisons are useful for project managers to analyse the performance of the project. It can be compared with other similar systems or the state-of-the-art, if there are, published values. Jones [109] has industry averages for various measures. Another example is the classical *make or buy* decision about complete systems or software components. If there are measurements possible for the different available components, they can be used to aid that decision.

#### Trend Analysis

Finally, there are cases in which we want not only a measure at a specific point in time but also its changes over time. This is somehow related to predictions, as we discussed with reliability growth models. They also use the change over time of occurred failures to predict the future. It is also a basic tool in *statistical process control* SPC [65]. It is analysed whether the analysed data varies in a "normal" span or if there are any unusual events. For example, the growth rate of defects or code size can be useful for project management. Another example is the number of anomalies reported by a bug pattern tool or the number of clones found by clone detection. There might not be enough resources to fix all these issues, but at least it should be made sure that the number is not increasing anymore. Again, there are various aggregation issues about determining the correct measure for a component from its subcomponents, for example.

### Aggregation Theory

Aggregation is not only a topic for software measures but in any area that needs to combine large data into smaller, more comprehensible or storable chunks. We describe the general theory of aggregation aggregators first and then discuss the state in software engineering.

#### General Theory

There is a large base of literature on aggregation in the area of soft computing [18, 31] where it is also called *information fusion*. They use aggregation operators in the construction and use of knowledge-based systems. "Aggregation functions, which are often called aggregation operators, are used to combine several inputs into a single representative value, which can be subsequently used for various purposes, such as ranking alternatives or combining logical rules. The range of available aggregation functions is very extensive, from classical means to fuzzy integrals, from triangular norms and conorms to purely data driven constructions" [17]. We mainly build on the variety of work in that area in the following to explain what is now considered the basics of aggregation. We mainly use the classification and notation from [53] and [18].

Informally, aggregation is the problem of combining $n$-tuples of elements belonging to a given set into a single element (often of the same set). In mathematical aggregation, this set can be, for example, the real numbers. Then an aggregation operator $A$ is a function that assigns an $y$ to any $n$-tuple $(x_1, x_2, \ldots, x_n)$:

$$A(x_1, x_2, \ldots, x_n) = y \tag{2.1}$$

From there on, the literature defines additional properties that are requirements for a function to be called an *aggregation operator*. These properties are not all compatible, however. Yet, there seem to exist some undisputed properties that must be satisfied. For simplification, the sets that aggregation operators are based on are usually defined as $[0, 1]$, i.e. the real numbers between 0 and 1. Other sets can be used, and by normalisation to this set, it can be shown that the function is an aggregation operator. Additionally, the following must hold:

$$A(x) = x \text{ identity when unary} \tag{2.2}$$

$$A(0, \ldots, 0) = 0 \wedge A(1, \ldots, 1) = 1 \text{ boundary conditions} \tag{2.3}$$

$$\forall x_i, y_i : x_i \leq y_i \Rightarrow$$

$$A(x_1, \ldots, x_n) \leq A(y_1, \ldots, y_n) \text{ monotonicity} \tag{2.4}$$

The first condition only is relevant for unary aggregation operators, i.e. the tuple that needs to be aggregated only has a single element. Then we expect the result of the aggregation to be that element. For example, the aggregation of the size in LOC of a single Java class should be the original size in LOC of that class.

The boundary conditions cover the extreme cases of the aggregation operator. For the minimum as input there must be the minimum as output and vice versa. For example, if we aggregate a set of ten modules, all with 0 LOC, we expect that the aggregation result is also 0 LOC.

Finally, we expect that an aggregation operator is monotone. If all values stayed the same or increased, we want the aggregation result also to increase or at least stay the same. This is interesting for trend analyses. If we aggregate again the size of modules to the system level and over time one of the modules increased in size (and none decreased), we want also the aggregation of the size to increase.

Apart from these three conditions, there is a variety of further properties that an aggregation operator can have. We only introduce three more that are relevant for aggregation operators of software measures.

The first condition that introduces a very basic classification of aggregation operators is *associativity*. An operator is associative if the results stay the same no matter in what packaging the results are computed. This has interesting effects on the implementation of the operator as associative operators are far easier to compute. Formally, for an associative aggregation operator $A_a$ the following holds:

$$A_a(x_1, x_2, x_3) = A_a(A_a(x_1, x_2), x_3) = A_a(x_1, A_a(x_2, x_3)) \tag{2.5}$$

The next interesting property is *symmetry*. This is also known as *commutativity* or *anonymity*. If an aggregation operator is symmetrical, the order of the input arguments has no influence on the results. For every permutation $\sigma$ of $1, 2, \ldots, n$ the operator $A_s$ must satisfy

$$A_s(x_{\sigma(1)}, x_{\sigma(2)}, \ldots, x_{\sigma(n)}) = A_s(x_1, x_2, \ldots, x_n) \tag{2.6}$$

The last property we look at because it holds for some of the operators relevant for software measures is *idempotence*. It is also known as *unanimity* or *agreement*. Idempotence means that if the input consists of only equal values, it is expected that the result is also this value.

$$A_i(x, x, \ldots, x) = x \tag{2.7}$$

An example for a an idempotent software measure is clone coverage (Sect. 4.4.2) which describes the probability that a randomly selected line of code is copied in the system. We can calculate the clone coverage for each individual module, and if we aggregate modules with the same clone coverage, we expect the aggregation result also to be this exact same clone coverage.

#### Software Engineering

Only few contributions to the theory of aggregation operators in software measurement have been made. The main basis we can build on is the assignment of specific aggregation operators (especially for the central tendency) to scale types. Representational theory has been suggested as basis for a classification of scale types for software measures [61]. The scales are classified into *nominal*, *ordinal*, *interval*, *ratio* and *absolute*. Nominal scales assign only symbolic values, for example, *red* and *green*. They are introduced in detail above.

This classification provides a first justification for which aggregation operators can be used for which classes of scales. For example, consider the measures of central tendency such as median or mean. To calculate the mean value of a nominal measure does not make any sense. For example, what is the mean of the names of the authors of modules in a software system? This is only part of the possible statistics that we can use as aggregation operators. Nevertheless, we will use this classification as part of the discussion on the suitability of aggregation aggregators.

### Aggregation Operators

#### Grouping

A very high-level aggregation is to define a set of groups, probably with a name each, and assign the inputs to the groups. This allows a very quick comprehension and easy communication about the results. Problematic is that the information loss is rather large.

### Rescaling

An often used technique to be able to comprehend the large amount of information provided by various measures is to change the scale type by grouping the individual values. This is usually done from higher scales, such as ratio scales, to ordinal or nominal scales. For example, we could define a threshold value for test coverage. Above the threshold, the group is *green*; below it, it is *red*. This is useful for all purposes apart from trend analysis where it can be applied only in a few cases. It is not idempotent in general and it depends on the specifics of the rescaling whether it is symmetrical.

### Cluster Analysis

Another, more sophisticated way, to find regularities in the input is cluster analysis. It is, in some sense, a refinement of the rescaling described above by finding the groups using clustering algorithms. The *K-means* [140] algorithm is a common example of such algorithms. It works with the idea that the input are points scattered over a plain and there is a distance measure that can express the space between the points. The algorithms then work out which points should fall into the same cluster. This aggregator is not associative and not idempotent.

### Central Tendency

The central tendency describes what colloquially is called the average. There are several aggregation operators that we can use for determining this average of an input. They depend on the scale type of the measures they are aggregating. All of them are not associative but idempotent.

The *mode* is the only way for analysing the central tendency for measures in a nominal scale. Intuitively, it gives the value that occurs most often in the input. Hence, for inputs with more than one maximum, the mode is not uniquely defined. If the result is then defined by the sequence of inputs, the mode is not symmetrical. The mode is useful for assessing the current state of a system and for comparisons of measures in a nominal scale. For $n_1, \ldots, n_k$ being the frequencies of the input values, the mode $M_m$ is defined as

$$M_m(x_1, \ldots, x_k) = x_j \Leftrightarrow n_j = \max(n_1, \ldots, n_k). \tag{2.8}$$

The *median* is the central tendency for measures in an ordinal scale. An ordinal scale allows to enforce an order on the values and hence a value that is in the middle can be found. The median ensures that at most 50 % of the values are smaller and at most 50 % are greater or equal. The median is useful for assessing the current state and comparisons. The median $M_{0.5}$ is defined as

$$M_{0.5}(x_1, \ldots, x_k) = \begin{cases} x_{((n+1)/2)} & \text{if } n \text{ is odd} \\ \frac{1}{2}(x_{(n/2)} + x_{(n/2+1)}) & \text{otherwise} \end{cases} \tag{2.9}$$

For measures in interval, ratio or absolute scale, the *mean* is defined. There are mainly three instances of means: arithmetic, geometric and harmonic mean. The arithmetic mean is what usually is considered as average. It can be used for assessing the current state, predictions and comparisons. The arithmetic mean $M_a$ is defined as follows:

$$M_a(x_1, \ldots, x_n) = \frac{1}{n} \sum_{i=1}^{n} x_i \tag{2.10}$$

We can use the geometric mean for trend analysis. It is necessary when measures are relative to another measure. For example, the growth rates of the size of several releases could be aggregated using the geometric mean. The geometric mean $M_g$ is defined as

$$M_g(x_1, \ldots, x_n) = \sqrt[n]{\prod_{i=1}^{n} x_i}. \tag{2.11}$$

As it uses the product, it actually has the absorbent element 0 [53]. Finally, the harmonic mean needs to be used when different sources are combined and hence weights need to harmonise the values. An example would be when, to analyse the reliability of a system, the fault densities of the components are weighted based on their average usage. Given the weights $w_i$ for all the inputs $x_i$, the harmonic mean $M_h$ is given by

$$M_h(x_1, \ldots, x_n) = \frac{w_1 + \ldots + w_n}{\frac{w_1}{x_1} + \ldots \frac{w_n}{x_n}}. \tag{2.12}$$

Dispersion

In contrast to the central tendency, the dispersion gives an impression about how scattered the inputs are over their base set. Hence, we look at extreme values and their deviation from the central tendency.

The *variation ratio* is given by the proportion of cases which are not the mode. This is the only way for nominal measures to have a measure of dispersion. It indicates whether the data is in balance. This is useful for hot spot identification. The variation ratio $V$ is defined again using $(n_1, \ldots, n_k)$ as the frequencies of $(x_1, \ldots, x_k)$ by

$$V(x_1, \ldots, x_k) = 1 - \frac{\max(n_1, \ldots, n_k)}{k}. \tag{2.13}$$

Very useful operators for various analysis situations are the *maximum and minimum* of a set of measures. They can be used with measures of any scale apart from nominal. They are useful for identifying hot spots and for comparisons. They both are associative and symmetrical. The maximum *max* and the minimum *min* are defined as follows:

$$\forall x_i . y \geq x_i \Rightarrow \max(x_1, \ldots, x_n) = y \tag{2.14}$$

$$\forall x_i . y \leq x_i \Rightarrow \min(x_1, \ldots, x_n) = y \tag{2.15}$$

The *range* is the standard tool for analysing the dispersion. Having defined the maximum and the minimum above, it is easy to compute. It is given by the highest value minus the lowest value in the input. It can be useful for assessing the current state and comparisons. This operator is neither idempotent nor associative. The range $R$ is simply defined as

$$R(x_1, \ldots, x_n) = \max(x_1, \ldots, x_n) - \min(x_1, \ldots, x_n) \tag{2.16}$$

The *median absolute deviation* (MAD) is useful for ordinal metrics. It is calculated as the average deviation of all values from the median. This again can be used for current state analyses and comparisons when the median is the most useful measure for the central tendency. The median absolute deviation is not associative but symmetrical. It is defined as

$$D(x_1, \ldots, x_n) = \frac{1}{n} \sum_{i=1}^{n} |x_i - M_{0.5}(x_1, \ldots, x_n)|. \tag{2.17}$$

For other scales, the most common dispersion measures are the *variance* and the *standard deviation*. It is always the best choice for analysing the dispersion when the mean is the best aggregator for the central tendency. The standard deviation has the advantage over the variance that it has the same unit as the measure to be aggregated. Hence, it is easier to interpret. We use it again for analysing the current state and for comparisons. It is not associative but symmetrical. The variance $S^2$ and the standard deviation $S$ are defined as follows:

$$S^2(x_1, \ldots, x_n) = \frac{1}{n} \sum_{i_1}^{n} (x_i - M_a(x_1, \ldots, x_n))^2 \tag{2.18}$$

$$S(x_1, \ldots, x_n) = \sqrt{S^2(x_1, \ldots, x_n)} \tag{2.19}$$

Concentration

A step further than central tendency and dispersion are concentration aggregators. They measure how strongly the values are clustered over the whole span of values.

For concentration measures to be available, the measures need to be at least in interval scale. The basis for these concentration measures is often the *Lorenz curve*. It describes the perfect equality of a distribution. The Lorenz curve itself is built from the ordered inputs but alone can only give a graphical hint of the concentration.

The *Gini coefficient* is a concentration measure that is defined by the deviation of the actual distribution from the perfect Lorenz curve. Its normalised version measures the concentration in a scale between 0 (no concentration) and 1 (complete concentration). This can be used for a more sophisticated analysis of hot spots. If the operator assumes that the input is preordered, the Gini coefficient is not symmetrical. It is also not associative. The normalised Gini aggregation operator $G_n$ where $(i)$ indicates the $i$th element in the ordered input is defined as

$$G_n = \frac{2\sum_{i=1}^{n} i x_{(i)} - (n+1)\sum_{i=1}^{n} x_{(i)}}{n\sum_{i=1}^{n} x_{(i)}} \tag{2.20}$$

Ratios and Indices

A further way to quantitatively describe aspects that can be used as aggregation operators are ratios and indices. They are binary operators and therefore relate two different measures. It is important that the results of such operators are usually without a dimension or unit. This is because they only describe a relation. Because they are all defined as ratios, they are not associative but usually they are symmetrical. The calculation is straightforward; therefore, we refrain from giving explicit equations for each aggregation operator. The differences in this operators do not lie in the different calculation but are conceptual differences.

*Fraction measures* relate subsets to their superset. An example would be *comment lines/total lines of code*. It is suitable for assessing the current state or to make comparisons.

*Relation measures* are measures of two different metrics that are not in a subset–superset relation. An example is *deleted lines/remaining lines*. This is again interesting for comparisons or for trend analyses.

*Indices* describe the relationship between the result of a metric and a base set measured at different points in time. This fits to trend analyses. For example, the *produced lines of code* each month in relation to the *lines produced on average* in the company are an index.

#### Tool Support

The aggregation operators presented above can easily be implemented and included in quality evaluation tools. We did that prototypically for the tool ConQAT[5] [48]

[5] http://www.conqat.org/.

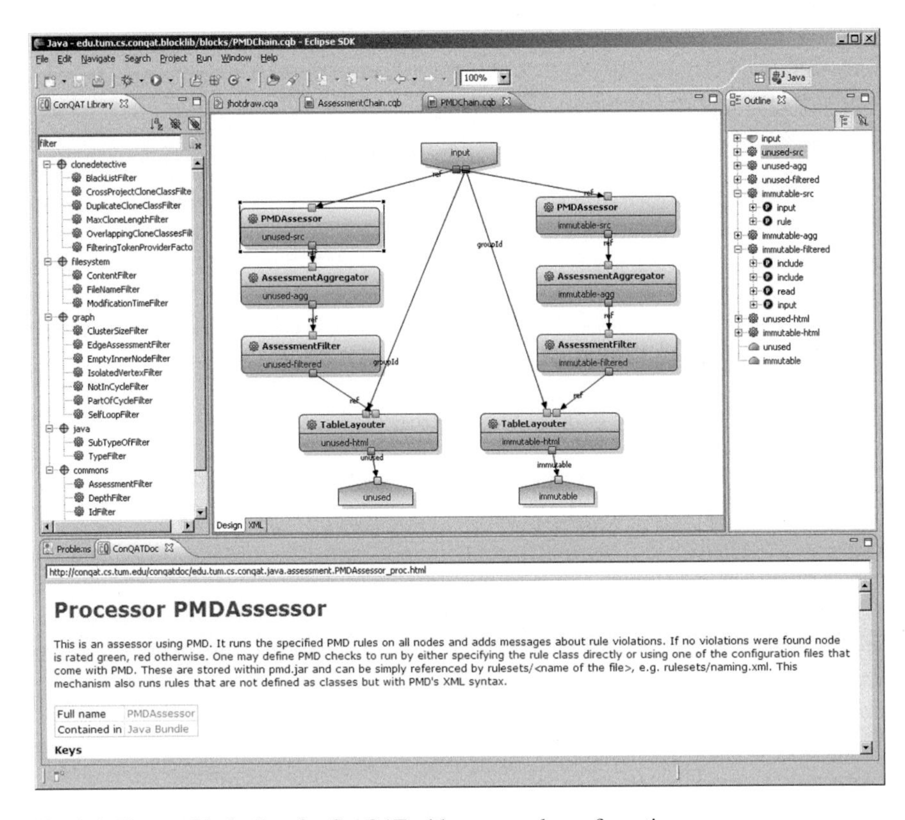


**Fig. 2.6** The graphical editor for ConQAT with an example configuration

which we also use as dashboard for continuous quality control. Furthermore, it implements the Quamoco quality evaluation approach (Sect. 4.2). It makes use of a pipes and filters architecture in which complex analyses can be built by connecting simple so-called *processors*. These processors read the artefacts of interest, perform analyses on them and write the resulting outputs. There is also a graphical editor in which the nets of processors can be modelled and run. An example is shown in Fig. 2.6.

ConQAT already provides various processors for different languages and artefacts, such as Java and C# source code, Matlab Simulink models or Bugzilla defect databases. It has been used in several industrial environments for a variety of analyses [48]. Aggregation is an important part of combining these processors.

The implementation of the aggregation operators can be done straightforwardly as a ConQAT processor each. The list of values to be aggregated is the input in the aggregator that calculates the result of the operator and writes it in the output. More interesting is the case when we operate on a hierarchy such as Java classes in packages. Then the aggregation can be done on each inner node in the hierarchy. Here the distinction between associative and non-associative aggregation operators becomes important. For associative operators, we only need to consider the direct children for each inner node. Non-associative operators require a complete traversal of the subtree.

It is not always the best way to implement processors for aggregators in the strict theoretical sense as described above. In theory, an aggregation operator always maps a list of values to a *single* value. Because of computational savings and convenience of use, however, it makes often sense to combine several aggregation operators in one processor. For example, the variation ratio is easily computed together with the mode. Actually, a separate implementation would require to identify the mode again. Also note that all the ratios and indices can be implemented in one processor as their differences are purely conceptual.

## Example

To demonstrate the use of the different aggregation operators and their respective advantages, we describe an example application to the open source project *Apache Tomcat*, a Java servlet container. The hypothetical context is a comprehensive quality evaluation of this software system. We briefly describe Tomcat, the investigated evaluation scenarios and the achieved results with the aggregation operators.

### Apache Tomcat

Tomcat[6] is a project by the Apache Software Foundation that is also responsible for the most successful web server. Tomcat is also aiming at the web but is a servlet container, i.e. able to run servlets and Java Server Pages (JSP). Several of the best developers in that area contribute to Tomcat as it is also used in the reference implementation of the corresponding specifications.

The version 6.0.16 that we used in this example conforms to the servlet specification 2.5 and the JSP specification 2.1. It contains various functionalities needed in a web and servlet context: network management, monitoring, logging and balancing.

### Analysis Scenarios

We perform a quality evaluation of Tomcat using the dashboard toolkit ConQAT including the implementation of the aggregation operators as described above. For the analysis, we need a set of scenarios that form the quality issues that are currently of interest. These span quite different aspects of the system and are aimed at showing the various possibilities of the aggregation operators. We concentrate on only three scenarios that illustrate the use of the aggregators:

[6]http://tomcat.apache.org/.

First, we analyse the *average class size*. Although a problematic measure in many aspects, *lines of code* (LOC) is reasonable to get an impression of the size of a system and of its components. To aid code reading and comprehension, it is advantageous not to have too large classes. Therefore, we want an analysis that shows us the average size of classes overall and for each package. This is purely for assessment as no direct actions can be derived from the results.

Second, we analyse the activity of the different developers by assessing who is the *most frequent author*. This can be interesting just for identifying and gratifying the most productive author. Although this should be done with care as only the authorship of classes is not a meaningful measure for productivity. If clear responsibilities for specific packages exist, however, these author aggregations can also be used to check whether the responsible persons actually work the most on their packages.

Third, one measure that is considered important in object-oriented development is the *coupling between classes* [34]. We are interested in how strongly the classes in the software are interconnected. For this, we want to know how many classes each class depends on in relation to the size of the class. The latter requirement comes from the fact that the dependence on other class can clearly be influenced by its size: a larger class potentially relies on more classes.

#### Results

The corresponding ConQAT configurations for the scenarios were developed by combining existing ConQAT processors with the aggregation operators as described above. Actually, we developed one configuration that wrote the results for the three scenarios in one HTML output.

Size analyses are a practical way to get a feeling for a system. The *average class size* is a suitable measure for that purpose when analysing Java software. For the analysis, we use the mean operator and the range operator. This way, we see not only the averages for the classes but also some information about the dispersion of the values. For illustration, the output of ConQAT is shown in Fig. 2.7. The package `org.apache.tomcat`, for example, has a mean class size of 174.724 LOC with the large range of 2,847 LOC. In contrast, the package `org.apache.catalina.tribes.transport.bio` has the higher mean of 181.111 LOC but only a range of 224 LOC. Hence, the sizes of the classes in the latter package are much more uniform.

We do not have a list of responsibilities for the packages of Tomcat. Hence, we only analyse which are the *most frequently found authors* in which packages. First of of all, it becomes obvious with this analysis whether the *author* tag from JavaDoc is consistently used in a project. In Tomcat it is not used in all classes but in many. As we only look at names, which are of a nominal scale, the mode operator has to be used. The application of the mode shows that actually the most frequent author is the empty one. More interesting are specific subpackages: the package `org.apache.catalina.connector` is dominated by Remy

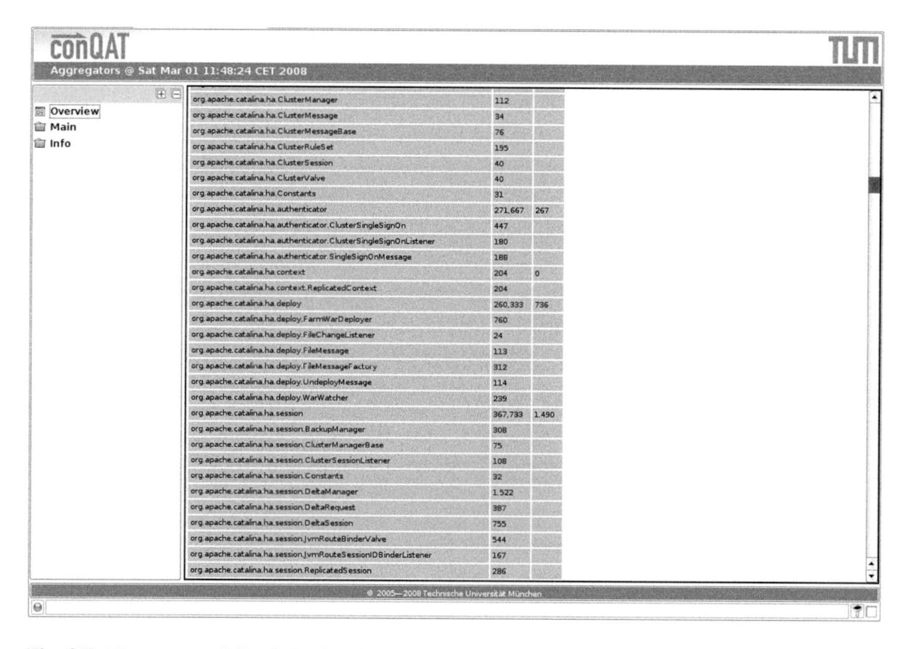


**Fig. 2.7** The output of ConQAT for the average class sizes

Maucherat or `org.apache.catalina.valves` by Craig R. McClanahan. To add more information to these statements, we also used the variation ratio operator that determines the fraction of all others that are not the mode. For the complete project, we learn that although the empty author is the most frequent one, 83.3 % of the classes have named authors. In the connector package Remy Maucherat is not the author of 76.5 % of the classes, and in the valves package Craig R. McClanahan is not the author of 60 % of the classes.

We restrict the *coupling of classes* analysis to the fan-in of each class. This measures the number of classes that depend on the current class. ConQAT already provides the processor *ClassFanInCounter* for that purpose. Running that processor on the Java code of Tomcat provides us with a corresponding list. Various aggregations of these values are possible. The maximum aggregator gives the class with the highest fan-in. For Tomcat, as probably in many systems, this is the logging class `Log` with 230 classes that depend on it. Also the mean or variation is an interesting aggregation to understand the distribution of the fan-ins. Finally, we can define a relation aggregator that measures the number of fan-ins per size in LOC. This gives an indication whether larger classes are more used than smaller classes as they might provide more functionality. However, this is not the case with Tomcat.

### Discussion

Our first observation is that the set of described aggregation operators is useful and sufficient in the encountered situations. Many questions can be answered with the most basic operators for central tendency and dispersion. Most of the time several of those basic operators are used in combination to get a better overview of the software characteristic under analysis.

Moreover, it is important to point out that the three basic properties of aggregation operators work well in categorising the operators. Several operators such as the minimum or the maximum are associative; others as the mean or the median are not. This is especially useful for implementation as the associativity allows a simpler and quicker implementation. In general, the associative and the non-associative operators were implemented with two different base classes. The idempotence holds for most of the operators but usually not for rescaling. Finally, the symmetry is a property of all the operators we discussed above which is also important for their implementation and use. The user of the operators never has to worry about in what sequence the input currently is.

**Summary**

We gave in this section a comprehensive but certainly not complete overview of software measures and measurement as well as a collection of useful aggregation operators for use in a variety of measurement systems. We judged the suitability of each operator based on a set of aggregation purposes because we see the purpose of the aggregation as the main driver of the operator choice. We also discussed further properties of operators such as associativity that are important for implementation and use of the operators. Moreover, the scale type of the measures is a further constraint that limits the available operators for a specific measure.

## 2.3 ISO/IEC 25010

After several years of development, a working group of the International Organization for Standardization released in 2011 a reworked software product quality model standard: ISO/IEC 25010 [97]. It is still strongly influenced by its predecessor ISO 9126 [107] but restructures and adds several parts of the quality models. This standard is likely to become the most well-known type of software quality models, and it will have an effect on existing quality models in practice. We therefore look into it in more detail.

### *2.3.1 Concepts and Meta-Model*

ISO/IEC 25010 grew historically from the initial hierarchical models from Boehm and McCall which were already the basis of the old ISO/IEC 9126 (cf. Sect. 2.1). The main modelling concept is hence to provide a taxonomy that breaks the complex

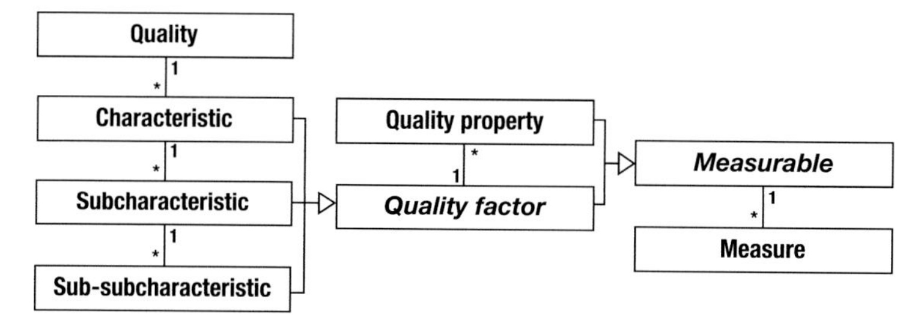


**Fig. 2.8** Implicit meta-model of ISO/IEC 25010

concept of software product quality into smaller, hopefully more manageable parts. The idea is that the decomposition reaches a level on which we can measure the parts and use that to evaluate the software product's quality.

A UML class diagram of the meta-model of the ISO/IEC 25010 quality models is shown in Fig. 2.8. It shows the hierarchical structure that divides quality into *characteristics*, which can consist of *subcharacteristics* and, in turn, of *sub-subcharacteristics*. We call all of these *quality factors* and they might be measurable by a *measure*. If a direct measurement is not possible, we use measurable *quality properties* that cover the quality factor. Please note that the terms *quality factor* and *measurable* are not part of the standard, but we introduced them here for the sake of a well-structured meta-model.

Using this hierarchical structure, the standard then assumes that we can build a quality model. ISO/IEC 25010 contains a quality in use model and a product quality model which we will cover in the next sections. In addition, there is a compliant data quality model in ISO/IEC 25012 [98]. These models shall help the software engineers as a kind of checklist so that they consider all relevant parts of software quality in their products. Nevertheless, the standard emphasises that not all characteristics are relevant in any kind of software. It gives no help how to customise the quality model, however.

The description of the quality model structure explicitly mentions the connection to measures. Hence, it is considered important to be able to quantify quality and its characteristics. The details of this is not part of the quality model, however. The standard ISO/IEC 25040 [102] adds quality evaluation to the models. We will discuss quality evaluation in more detail in Sect. 4.2.

## 2.3.2 Product Quality Model

We have used this quality model already in the introduction to discuss the various areas of software quality (Sect. 1.3.3). Furthermore, the product quality model fits best to traditional software product quality models that we discussed in Sect. 2.1. It describes the most common "-ilities" for software and several subcharacteristics for each. There is an overview in Fig. 2.9. The idea is that each of the characteristics

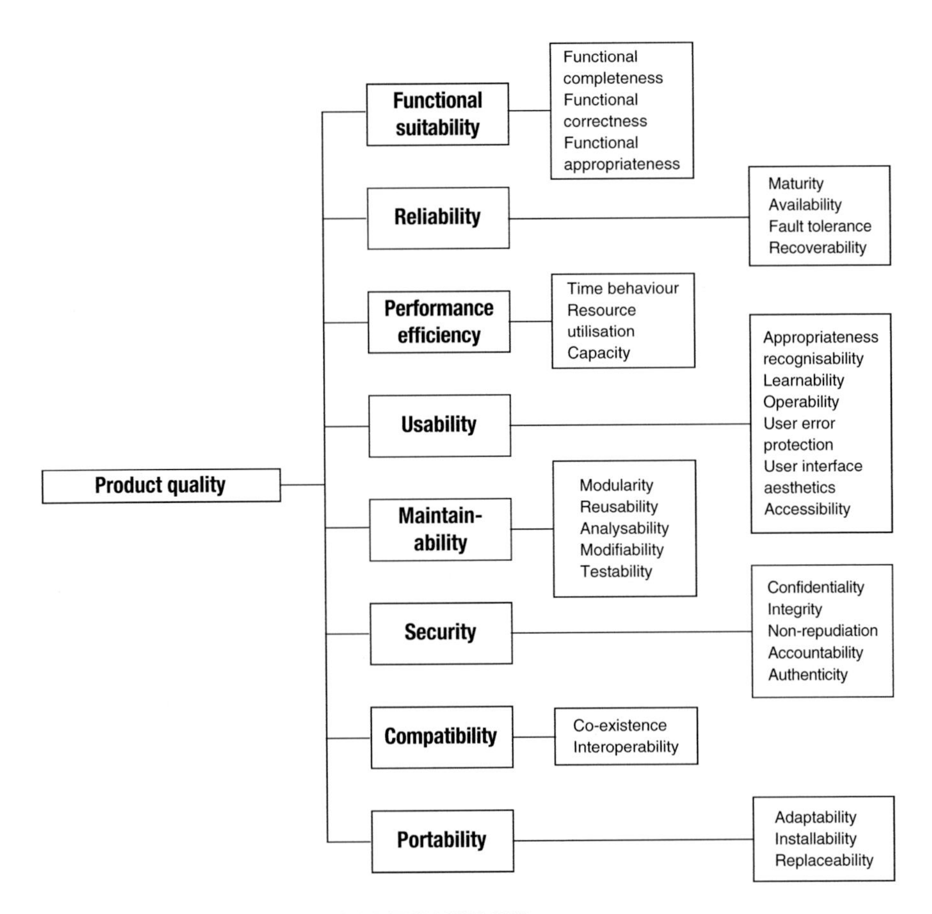


**Fig. 2.9** Product quality model of ISO/IEC 25010 [97]

is something we can analyse directly at the software product. In addition, the standard claims that they should also be valid for system quality.

The intention of the standard is that this list of eight characteristics comprehensively describes the quality of a software product. *Functional suitability* means that the product fits to the functional needs and requirements of the user and customer. *Performance efficiency* describes how well the product responds to user requests and how efficient it is in its execution. Today's software products rarely operate in an isolated environment and therefore *compatibility* defines the quality that a product does not disturb or can even work together with other products. The characteristic *usability* subsumes the aspects of how easy the system can be used. This includes how fast using it can be learned but also if the interface is attractive and if challenged persons are able to use it.

Many people mean *reliability* when they talk about quality. Reliability problems are – together with performance efficiency – usually the most direct problems of users. The product produces failures and hence might not be available enough for

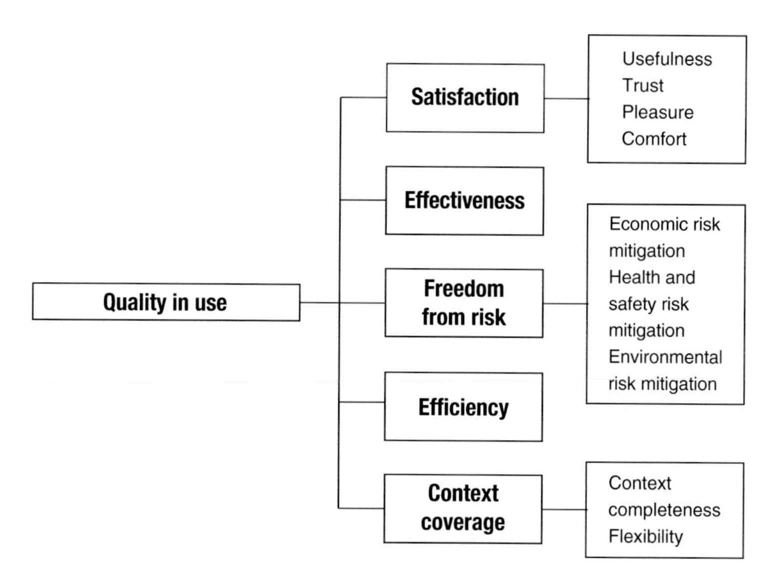


**Fig. 2.10** Quality in use model of ISO/IEC 25010 [97]

the user. Of increasing importance is *security* as most systems are now accessible over networks. It includes to keep data intact and a secret as well as that the product needs to make sure that its users are who they claim to be. A characteristic aimed towards the developers is *maintainability* which describes that the system should be designed and programmed in a way that it is easy to comprehend, change and test. Finally, *portability* is also most important for developers who need to bring the product to another platform (programming language, operating system, hardware). It means that the necessary changes can be easily done and it can be easily installed.

### 2.3.3 Quality in Use Model

While the product quality model wants to describe the characteristics of the product directly, the quality in use model looks at characteristics of the interactions of different stakeholders with the product. The most prominent of those stakeholders is the primary user. This is why quality in use is often associated merely with usability. Quality in use, however, can also mean the quality in maintaining or porting the product or providing content for the product. Hence, to "use" and "user" have a very broad meaning in this context. The model describes the quality in use with five characteristics shown in Fig. 2.10.

The two core characteristics of quality in use are effectiveness and efficiency. *Effectiveness* is how well the product supports the user in achieving objectives, and *efficiency* denotes the amount of resources necessary to achieve these objectives. For drivers interacting with navigation systems in cars, it means: Can they get suitable navigations to their desired destinations and how many times do they have to press a button for that? For maintainers, this is rather a maintenance task characterised by the number of new faults introduced and effort spent.

Quality in use goes beyond reaching your objectives with appropriate resources. The *satisfaction* of the user is also a quality characteristic that includes the trust the product inspires the users with and also the pleasure of using the product. In many contexts, also the *freedom from risk* is important. The most famous part thereof is the safety of systems that can harm humans. Environmental or economic risks play a role here as well. Finally, the standard also proposes that the usage should cover the appropriate context. *Context coverage* contains that the product understands the complete context of its usage and it can react flexible to changes in the context.

### 2.3.4 Summary and Appraisal

ISO/IEC 25010 provides with its two quality models a comprehensive list of relevant quality characteristics for software products and the interaction with these products. It is a useful structuring of the relevant topics and it can serve as a good checklist to help requirements engineers in not forgetting any quality characteristics and also quality engineers to analyse the quality of a system. In that, it is an improvement over the old ISO/IEC 9126, because it made some useful changes in the models, such as including security as a separate characteristic. It also reduced the number of quality models from three to two. It is still not clear, however, when it is advisable to use which model. It seems that most software companies employ only the product quality model and only when they analyse usability, the quality in use model is taken into account. The standard does not prescribe when to use which model apart from saying "The product quality model focuses on the target computer system that includes the target software product, and the quality in use model focuses on the whole human-computer system that includes the target computer system and target software product" [97]. It is important to understand, however, that these models are taxonomies. They describe one possible structuring of quality but in no way the only possible or sensible structuring. There will always be discussions about why a particular subcharacteristic is part of one characteristic and not the other. For example, is availability a subcharacteristic of reliability or security? In the end, for being a checklist, this does not matter. If you want to use the quality models beyond that, this simple structuring might not be sufficient. We will look into a more generic and more complete quality modelling framework in the next section.

## 2.4 Quamoco Quality Models

Already before the SQuaRE series and in parallel to its development, there had been research into improving ISO/IEC 9126 as the standard means to model and evaluate software product quality. Various research groups and companies worked on quality models and how they can be applied best. One of the larger initiatives was the three-year German project *Quamoco*[7] sponsored by the German ministry for research and education. It brought together experts from industry (Capgemini, itestra, SAP AG and Siemens) and research (Fraunhofer IESE and Technische Universität München). After it had finished at the beginning of 2012, it delivered generic quality model concepts, a broad base model and comprehensive tool support all available under an open source licence [211]. It is more generic than ISO/IEC 25010, but it can be used in a way conforming to the ISO standard. We look in detail into the core results in the following.

### 2.4.1 Usage of Quality Models

Most commonly, we find quality models reduced to mere reference taxonomies or implicitly implemented in tools (see Sect. 2.1). As explicit and living artefacts, however, they can capture general knowledge about software quality, accumulate knowledge from their application in projects and allow defining a common understanding of quality in a specific context [51, 82, 138, 143].

In contrast to other quality models that are expressed in terms of prose and graphics only, our quality model is integrated into the software development process as basis of all quality assurance activities. As Fig. 2.11 shows, the model can be seen as project- or company-wide quality knowledge base that centrally stores the definition of quality in a given context. Experienced quality engineers are still needed for building the quality models and enforcing them with manual review activities. They can rely on a single definition of quality, however, and are supported by the automatic generation of guidelines. Moreover, quality assessment tools like static analysers, which automatically assess artefacts, and test cases can be directly linked to the quality model and do not operate isolated from the centrally stored definition of quality. Consequently, the quality profiles generated by them are tailored to match the quality requirements defined based on the model. We refer to this approach as *model-based quality control*.

The execution of this approach should be continuously in a loop. In the quality control loop [48], which we will discuss in detail in Sect. 4.1, the quality model is the central element for identifying quality requirements, planning quality assurance, assessing the quality requirements using the quality assurance results and reworking the software product based on the assessment results. The quality model is useful for defining what we need to measure and how we can interpret it to understand the state of the quality of a specific product. A single source of quality information avoids redundancies and inconsistencies in various quality specifications and guidelines. Furthermore, if there is a certification authority, we can also certify a product against

[7] http://www.quamoco.de.

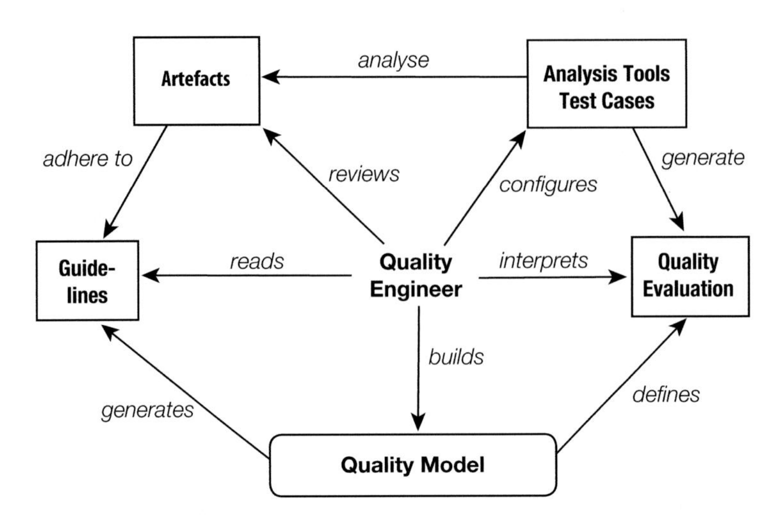


**Fig. 2.11** Model-based quality control [46]

the quality model. This would be useful, for example, if we included guidelines from standards such as ISO 26262 [105] or Common Criteria [43].

On top of that, the model itself helps us to establish suitable and concrete quality requirements. The quality model contains quality knowledge that we need to tailor to the product to be developed. This includes removing unneeded quality characteristics as well as adding new or specific quality factors.

## 2.4.2 Concepts

The previous work of all Quamoco partners on quality models, our joint discussions and experiences with earlier versions of the meta-model brought us to the basic concept of a *factor*. A factor expresses a *property* of an *entity* which is similar to what Dromey [54] calls *quality carrying properties* of *product components*. We use *entities* to describe the things that are important for quality and *properties* for the attributes of the things we are interested in. Because the concept of a factor is general, we can use it on different levels of abstraction. We have concrete factors such as *cohesion of classes* as well as abstract factors such as *portability of the product*. To illustrate the following concepts, we show their relationships in Fig. 2.12.

To clearly describe quality from an abstract level down to concrete measurements, we differentiate the two factor types *quality aspects* and *product factors*. Both can be refined into sub-aspects and sub-factors, respectively. The quality aspects express abstract quality goals, for example, the quality characteristics of ISO/IEC 9126 and ISO/IEC 25010 which would have the complete product as their entity. The product factors are measurable attributes of parts of the product. We require the leaf product factors to be concrete enough to be measured.

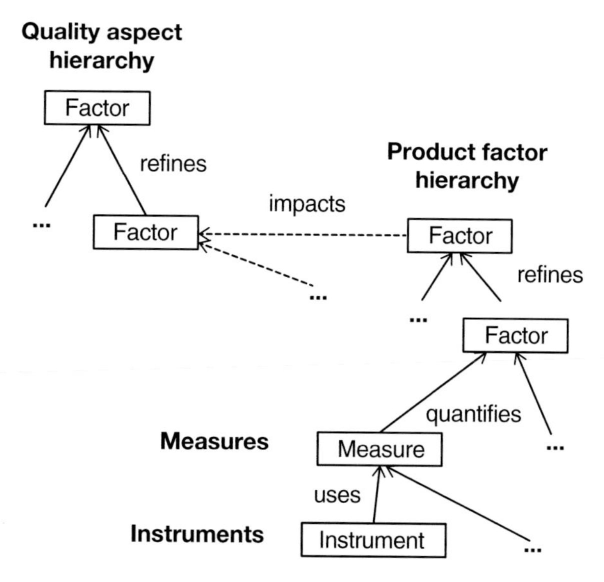


**Fig. 2.12** The Quamoco quality model concepts

An example is the *duplication of source code* which we measure with *clone coverage*[8] (see Sect. 4.4.2). This clear separation helps us to bridge the gap between the abstract notions of quality and concrete implementations. In addition, separating the entities from their properties addresses the problem of the difficult decomposition of quality characteristics for product factors. For example, the entity *class* can be decomposed straightforwardly into its attributes and methods. Furthermore, the entity concept is abstract enough to model processes or people as entities which could also have quality-carrying properties.

Moreover, we are able to model several different hierarchies of quality aspects to express divergent views on quality. Quality has so many different facets that a single quality factor hierarchy is not able to express it. Also in ISO/IEC 25010, there are two quality hierarchies: product quality and quality in use. We can model both as quality aspect hierarchies, and other types of quality aspects are also possible. For example, we experimented with activity-based quality models [51] (similar to quality in use of ISO 25010) and technical classifications [175]. We found that quality aspects give us the flexibility to build quality models tailored for different stakeholders. In principle, the concept also allows us to have different product factor hierarchies or more levels of abstraction. In our experiences with building quality models, however, we found the two levels *quality aspect* and *product factor* to be sufficient.

To completely close the gap between abstract quality characteristics and assessments, we need to put the two factor types into relation. The product factors have *impacts* on quality aspects. This is similar to variation factors which have impacts on

[8]Clone coverage is the probability that a randomly chosen line of code is duplicated.

quality factors in GQM abstraction sheets [191]. An impact is positive or negative and describes how the degree of presence or absence of a product factor impacts a quality aspect. This gives us a complete chain from measured product factors to impacted quality aspects, and vice versa.

We need product factors that are concrete enough to be measured so that we can close the abstraction gap. Hence, we have the concept of *measures* for product factors. A measure is a concrete description of how a specific product factor should be quantified for a specific context. For example, this could be the number of deviations of a rule for Java such as that strings should not be compared by "==". A factor can have more than one measure if we need multiple measures to cover the concept of the product factor.

Moreover, we separate the measures from their *instruments*. The instruments describe a concrete implementation of a measure. In the example of the string comparison, an instrument is the corresponding rule as implemented in the static analysis tool *FindBugs*. This gives us additional flexibility to collect data for measures manually or with different tools in different contexts. Overall, the concept of a measure also contributes to closing the gap between abstract quality factors and concrete software, as there is traceability from quality aspects via product factors to measures and instruments.

Having these relationships with measures and instruments, it is straightforward to assign evaluations to factors so that we can aggregate from the measurement results (provided by the instruments) to a complete quality evaluation. There are different possibilities for implementing this. We will describe a detailed quality evaluation method using these concepts in Sect. 4.2. Moreover, we can go the other way round. We can pick quality aspects, for example, ISO/IEC 25010 quality characteristics that we consider important and costly for a specific software system and trace what product factors affect them and what measures quantify them [203]. This allows us to concentrate on the product factors with the largest impact on these quality aspects. It also gives us the basis for specifying quality requirements, what we will discuss in detail in Sect. 3.1.

Building quality models in such detail results in large models with hundreds of model elements. Not all elements are important in every context and it is impractical to build a single quality model that contains all measures for all relevant technologies. Therefore, we introduced a modularisation concept which allows us to split the quality model into *modules*. Modularisation enables us to choose appropriate modules and extend the quality model by additional modules for a given context. The *root* module contains general quality aspect hierarchies as well as basic product factors and measures. In additional modules, we extend the root module for specific technologies and paradigms, such as object orientation; programming languages, such as C#; and domains, such as embedded systems. This gives us a flexible way to build large and concrete quality models that fit together, meaning they are based on the same properties or entities. As basis for all specialised quality models, we built the *base model*, a broad quality model with the most common and important factors and measures, to be applicable to (almost) any software. We will describe the base model in more detail in Sect. 2.4.6.

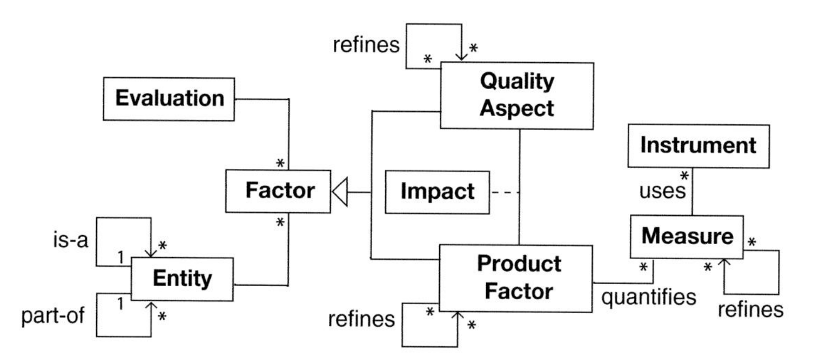


**Fig. 2.13** The Quamoco quality meta-model

## *2.4.3 Meta-Model*

We specified the general concepts described so far in a meta-model to make it precise. The benefit of an explicit meta-model is twofold: First, it ensures a consistent structure of quality models. Second, it constitutes a necessary basis for modelling tool support. The core elements of the meta-model are depicted as an (abstracted) UML class diagram in Fig. 2.13. Please note that we left out a lot of details such as the IDs, names and descriptions of each element to make the diagram more comprehensible. The central element of the meta-model is the *Factor* with its specialisations *Quality Aspect* and *Product Factor*. Both can be refined and, hence, produce separate directed acyclic graphs. An *Impact* can only exist between a *Product Factor* and a *Quality Aspect*. This represents our main relationship between factors and allows us to specify the core quality concepts.

A *Factor* always has an associated *Entity* which can be in an is-a as well as a part-of hierarchy. An Entity of a Factor can, in principle, be any thing, animate or inanimate, that can have an influence on software quality, e.g. the source code of a method or the involved testers. As we concentrate on product quality, we only choose product parts as entities for the product factors. We further characterise these entities by properties such as STRUCTUREDNESS or CONFORMITY to form a product factor. For the quality aspects, we often use the complete product as the entity to denote that they usually do not concentrate on parts of the product. The property of an *Entity* that the *Factor* describes is expressed in the *Factor*'s name.

Each *Factor* also has an associated *Evaluation*. It specifies how to evaluate the *Factor*. For that, we can use the evaluation results from sub-factors or – in the case of a *Product Factor* – the values of associated *Measures*. A *Measure* can be associated with more than one *Product Factor* and has potentially several instruments that allow us to collect a value for the measure in different contexts, e.g. with a manual inspection or a static analysis tool.

We modelled this meta-model with all details as an EMF[9] model which then served as the basis for a quality model editor (see Sect. 2.4.8).

### 2.4.4 Product Entities and Product Factors

The major innovation and also the most useful new construct in the Quamoco quality models is the product factor. It is the centre of the quality model as it is measured and describes impacts on quality aspects and, thereby, the product's quality overall. Product factors are far more tangible than quality aspects as they always describe a concrete product part, an entity. This relationship to entities also allows us an hierarchical decomposition of product factors.

The organisation of entities in a hierarchy is mostly straightforward, because entities often already have hierarchical relationships. For example, methods are part of a class or an assignment is a kind of statement. In principle, there could be more complex relationships instead of hierarchies, but modelling and finding information tends to be easier if such complex relationships are mapped to hierarchies.

The top level is the complete Product which represents the root for all entities of the system. In principle, Quamoco quality models can describe more than the product. As we concentrate on product quality in this book, however, we focus on product entities only. In the example in Fig. 2.14, the product contains the three artefacts: source code, its user interface and its requirements. Again, these entities need to be further refined. For example, the source code could consist of functions as well as procedures – depending on the programming language.

The associations between entities in the entity hierarchy can have one of two different meanings: Either an entity is a part or a kind of its super-entity. Along the inheritance associations, properties could also be inherited. In our current tool support, however, we do not support inheritance because it made the relationships in the model more complex. In principle, the inheritance would allows us a more compact description and prevents omissions in the model. For example, naming conventions are valid for all identifiers no matter whether they are class names, file names or variable names. We made some experiences with that (Sect. 5.2) earlier and will look into including it again in the future.

A product factor uses these entities and describes their properties. In essence, we model the product factors similar as in an hierarchical quality model (Sect. 2.1). In addition, to make the quality models useful in practice, we always want concrete product factors that we can measure and evaluate. The granularity of the factors shown in the diagrams above is too coarse to be evaluated. Therefore, we follow the hierarchical approach in the product factors and break down high level factors into detailed, tangible ones which we call *atomic* factors. An atomic factor is a factor that

[9]Eclipse Modeling Framework, http://emf.eclipse.org/.

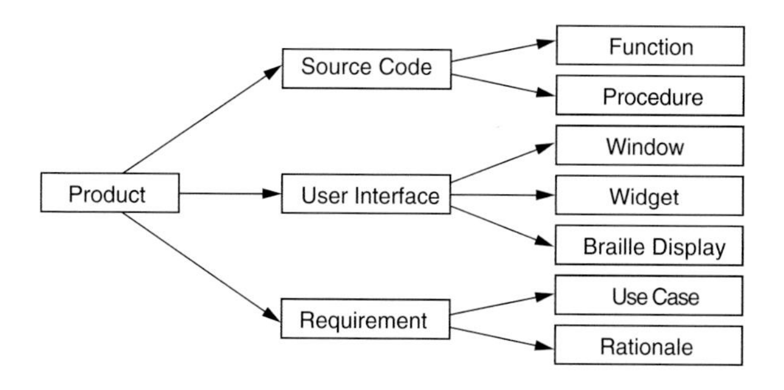


**Fig. 2.14** Example entities

can or must be evaluated without further decomposition either because its evaluation is obvious or there is no known decomposition.

We found that a fine-granular decomposition of the product factors inevitably leads to a high number of repetitions as the same properties apply to different kinds of artefacts. For example, *consistency* is required for identifier names as well as for the layout of the documentation. This is the reason, we explicitly introduced properties such as STRUCTUREDNESS or CONFORMITY to characterise entities. The difference is that entities "are the objects we observe in the real world" and properties are "the properties that an entity possesses" [123]. The combination of an entity and a property forms a *product factor*. We use the notation (introduced in [46]) [Entity | PROPERTY] when we need to refer to a product factor in the text. For the example of code clones, we write [Method | REDUNDANCY] to denote methods that are redundant.

An influence of a product factor is specified by an impact which can be positive or negative. Using the notation introduced for product factors, we can express the impact a product factor has on a quality aspect with a three-valued scale where "+" expresses a positive and "−" a negative impact (the non-impact is usually not made explicit):

$$\text{[Entity e | PROPERTY P]} \xrightarrow{+/-} \text{[Quality Aspect a]}$$

Examples are [Debugger | EXISTENCE] $\xrightarrow{+}$ [Fault Diagnostics] which describes that the existence of a debugger has a positive influence on the activity fault diagnostics. [Identifier | CONSISTENCY] $\xrightarrow{+}$ [Analysability] describes that consistently used identifier names have a positive impact on the analysability of the product. As another example, [Variable | SUPERFLUOUSNESS] $\xrightarrow{-}$ [Code Reading] denotes that unused variables hamper the reading of the code. To overcome the problem of unjustified quality guidelines and checklists each impact is additionally equipped with a detailed description.

Note that the separation of entities and properties does not only reduce redundancy but allows us a clear refinement of the product factors. Let us illustrate that by an example of the quality taxonomy defined in [168]: *System Complexity*. As *System Complexity* appears too coarse-grained to be assessed directly, it is desirable to further decompose this element. The decomposition is difficult, however, as the decomposition criterion is not clearly defined, i.e. it is not clear what a subelement of *System Complexity* is. A separation of the entity and the property as in [System | COMPLEXITY] allows for a cleaner decomposition as entities themselves are not valued and can be broken up in a straightforward manner, e.g. in [Subsystem | COMPLEXITY] or [Class | COMPLEXITY].

## 2.4.5 Measures and Instruments

For the purpose of defining quality, product factors, quality aspects and impacts between them are sufficient. Often, however, we want to grasp quality more concretely and prescribe detailed quality requirements or evaluate the quality of a software system. Then we specify measures that evaluate each atomic factor either by automatic measurement or by manual review. For higher-level statements about the quality of a system, we need to aggregate the results for these atomic measurements. The aggregation can be done along the refinement hierarchies of the product factors, quality aspects, and along the impacts. How you exactly perform the aggregation depends on your evaluation goal. We will discuss a detailed and tool-supported evaluation method in Sect. 4.2.

The product factors are the core elements of the model that need to be evaluated to determine the quality of a product. For this evaluation, we provide explicit measures for each factor. Since many important factors are semantic in nature and therefore not assessable in an automatic manner, we distinguish three measure categories:

1. Measures that we can collect with a tool. An example is an automated check for `switch` statements without a `default` case which could measure the product factor [Switch Statement | COMPLETENESS].
2. Measures that require manual activities, e.g. reviews. An example is a review activity that identifies the location and number of improper use of data structures for the corresponding [Data Structures | APPROPRIATENESS].
3. The combination of measures that we can assess automatically to a limited extent requiring additional manual inspection. An example is redundancy analysis where we detected cloned source code with a tool, but other kinds of redundancy must be left to manual inspection ([Source Code | REDUNDANCY]).

When we include measures in the quality model, it becomes similar to the result of the Goal-Question-Metric (GQM) approach. Another way of looking at Quamoco quality models is therefore as GQM patterns [12, 135]. The quality aspect defines the goal and the product factors are questions for that goal which are measured by certain metrics in a defined evaluation. For example, the goal is to evaluate modifiability which is analysed by asking the question "How consistent are the identifiers?" which in turn is part of an inspection.

We further distinguish *instruments* from *measures*. Measures are the conceptual description of a measurement, while an instrument is the concrete way for collecting data for that measure. For example, the product factor about complete switch statements [Switch Statement | COMPLETENESS] has a measure *Missing default case* which counts the number of switch statements without a default case. For Java, we have an automatic instrument: The static analysis tool FindBugs has the rule *SF_SWITCH_NO_DEFAULT* that returns findings suitable for the measure. This distinction allows us to describe measures to some degree independent of the used programming language or technology and to provide different instruments depending on the available tools. For example, if no tools are used, the switch statements could also be checked in a manual review which would be a manual instrument.

### 2.4.6 Quality Aspects: Product Quality Attributes

The traditional and well-established way to describe quality is using a quality taxonomy of quality attributes or quality characteristics. As discussed earlier, this approach goes back to the first quality models of Boehm et al. [25] and McCall and Walters [148] from the 1970s. The most recent incarnation of this approach is defined in the product quality model of ISO/IEC 25010 (see Sect. 2.3). Therefore, the most straightforward quality aspect hierarchy is to directly use the quality characteristics defined there. The quality characteristics of ISO/IEC 25010 become quality aspects, and instead of assigning directly measure elements to the quality characteristics, as defined in the standard, we specify measurable product factors with impacts onto those quality characteristics. We can see that as an implementation of the quality property concept of the standard.

Therefore, we have quality aspects such as *reliability*, *usability* or *maintainability*. They have the corresponding sub-aspects as they have quality sub-characteristics in ISO/IEC 25010: The sub-aspects of *reliability* are *maturity*, *availability*, *fault tolerance* and *recoverability*. The sub-aspects of *maintainability* are *modularity*, *reusability*, *analysability*, *modifiability* and *testability*. It is a simple transformation from the standard to the Quamoco meta-model. For all quality aspects, the corresponding entity is Product. Hence, the top-level quality aspect is [Product | QUALITY] for this quality aspect hierarchy. As the main information for this quality aspect hierarchy lies in the properties, i.e. the quality characteristics and attributes, we often omit Product as an abbreviation.

Let us look at three examples of product factors and their influence on this kind of quality aspects. The product factor [Subroutine | TERMINATION] describes "A method terminates if its execution ends in a finite time correctly". The product factor is measured by analyses of infinite recursions and infinite loops. If the product factor is present, it has a positive influence on the reliability of the product:

[Subroutine | TERMINATION] $\xrightarrow{+}$ [Reliability]

The product factor [Boolean expression | UNNECESSARILY COMPLICATED] is defined as "A boolean expression is unnecessarily complicated if the intended purpose could be met easier". This product factor is a refinement of the product factor [Source code part | UNNECESSARILY COMPLICATED]. This means that there are different source code

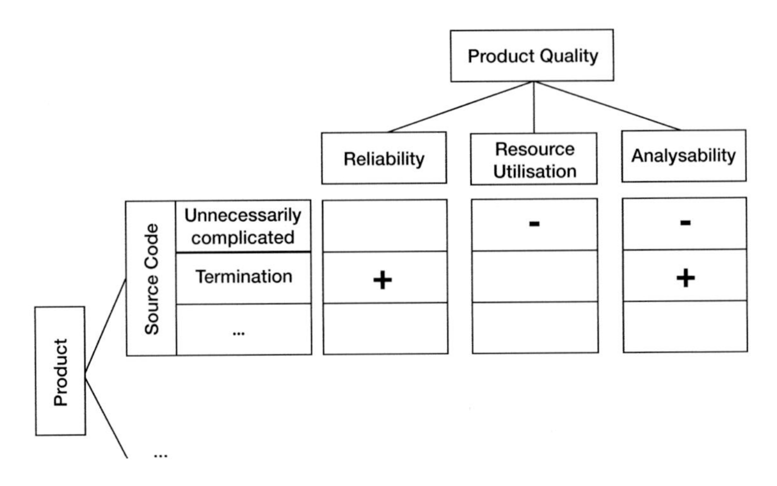


**Fig. 2.15** Quality matrix for product quality attributes

parts that can be unnecessarily complicated in different ways. In this case, we have an influence on resource utilisation, which is a sub-aspect of performance efficiency, as the Boolean expression needs more memory than actually necessary:
[Boolean expression | UNNECESSARILY COMPLICATED] $\xrightarrow{-}$ [Performance efficiency]

The same product factor influences also analysability, which is a sub-aspect of maintainability, because the code is more complex and, hence, harder to understand:
[Boolean expression | UNNECESSARILY COMPLICATED] $\xrightarrow{-}$ [Analysability]

Figure 2.15 shows an excerpt of the quality model using product quality characteristics as quality aspects visualised as a matrix. It depicts the decomposition of the product into the source code and corresponding product factors. The "+" and "–" show a positive or negative impact on a part of the quality aspect hierarchy.

### *2.4.7 Quality Aspects: Activity Based or Quality in Use*

A way to make the abstract and multifaceted concept of quality more concrete is to map it to costs: the cost of quality (Sect. 1.3). Yet, it is also difficult to find and quantify these costs. We were inspired by the concept of *activity-based costing* that uses activities in an organisation to categorise costs. It enables us to reduce the large category of indirect costs in conventional costing models by assigning costs to activities. Why should this assigning of costs to activities not work for quality costs? As a consequence, we introduced activities as first-class citizens into quality models [46, 51].

Besides, we were also motivated from our experiences with building large hierarchical quality models for maintainability. In this process it became harder and harder to maintain a consistent model that adequately describes the interdependencies

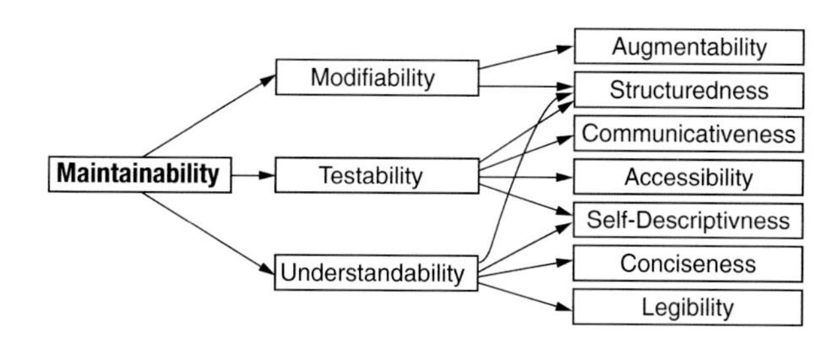


**Fig. 2.16** Software Quality Characteristics Tree

between the various quality criteria. A thorough analysis of this phenomenon revealed that our model, and indeed most previous models, mixed nodes of two different kinds: maintenance *activities* and *characteristics* of the system to maintain. An example of this problem is the *maintainability* branch of Boehm's *Software Quality Characteristics Tree* [25] in Fig. 2.16.

Though (substantivated) adjectives are used as descriptions, some nodes refer to activities (modify, test, understand), whereas others describe system characteristics (structuredness, self-descriptiveness), albeit very general ones. So the model should rather read as: When we *maintain* a system we need to *modify* it and this activity of *modification* is (in some way) influenced by the *structuredness* of the system. While this difference may not look important at first sight, we found that this mixture of activities and characteristics is at the root of the decomposition problem encountered with previous models. The semantics of the edges of the tree is unclear or at least ambiguous because of this mixture. Since the edges do not have a clear meaning, they neither indicate a sound explanation for the relation of two nodes nor can we use them straightforwardly to aggregate values.

As the actual maintenance efforts strongly depend on both, the type of system and the kind of maintenance activity, it is necessary to distinguish between activities and characteristics. For example, consider two development organisations of which company $A$ is responsible for adding functionality to a system while company $B$'s task is merely fixing bugs of the same system just before its phase-out. One can imagine that the success of company $A$ depends on different quality criteria (e.g. architectural characteristics) than company $B$'s (e.g. a well-kept bug tracking system). While both organisations will pay attention to some common characteristics such as a good documentation, $A$ and $B$ would and should rate the maintainability of the system in quite different ways, because they are involved in fundamentally different *activities*. Looking at maintainability as the productivity of the maintenance activity within a certain context widens the scope of the relevant criteria. $A$ and $B$'s productivity is determined not only by the system itself but also by a plethora of other factors, which include the skills of the engineers, the presence of appropriate software processes and the availability of proper tools like debuggers. Therefore, our quality models are in principle not limited to the software system itself, but we can describe the whole *situation* [51]. In this book, however, we focus on product quality and, hence, will consider only *product factors*.

Based on these insights, we proposed activity-based quality models (ABQM) to address the shortcomings of existing quality models [51]. They use the idea of avoiding high-level "-ilities" for defining quality and instead break quality down into detailed factors and their influence on activities performed on and with the system. Over time, we have found that ABQMs are applicable not only to maintainability but also to all quality factors. We have built concrete models for maintainability [51], usability [217] and security (see Sect. 2.6.2). The basic elements of ABQMs were also product factors – characteristics of the system – together with a justified impact on an activity.

Therefore, we can use activities as a quality aspect hierarchy. An entity of an activity as quality aspect can be any activity performed on and with the system such as *modification* or *attack*. For activities, we often omit an explicit property and assume we mean effectiveness and efficiency of the activity. This also fits well with the quality in use model of ISO/IEC 25010 that describes different usages of the system where *use* can include, for example, maintenance. Hence, uses from ISO/IEC 25010 are equivalent to activities from ABQMs.

Let us consider the example of a web shop written in Java. If there are redundant methods in the source code, also called clones (see Sect. 4.4.2), they exhibit a negative influence on *modifications* of the system, because changes to clones have to be performed in several places in the source code:

[Method | REDUNDANCY] $\xrightarrow{-}$ [Modification]

Another important activity is the *usage* of the system which is, for example, influenced by empty *catch* blocks in the code which show that exceptions are not handled:

[Catch Block | EMPTINESS] $\xrightarrow{-}$ [Modification]

We can again visualise the separation of quality aspects and product factors as a two-dimensional quality model. As top level, we use the activity Activity which has sub-activities such as Use, Maintenance or Administration. These examples are depicted in Fig. 2.17. The product hierarchy is shown on the left and the activity hierarchy on the top. The impacts are entries in the matrix where a "+" denotes a positive and a "–" a negative impact.

The matrix points out what activities are affected by which product factors and allows us to aggregate results from the atomic level onto higher levels in both hierarchies because of the unambiguous semantics of the edges. Hence, one can determine that concept location is affected by the names of identifiers and the presence of a debugger. Vice versa, cloned code has an impact on two maintenance activities. The example depicted here uses a Boolean relation between factors and activities and therefore merely expresses the existence of a relation between a factor and an activity. To express different directions and strengths of the relations, you can use more elaborate scales here. For quality evaluations, we can further refine this to a numerical scale (see Sect. 4.2).

The aggregation within the two hierarchies provides a simple means to cross-check the integrity of the model. For example, the sample model in Fig. 2.17 states that tools do not have an impact on coding which is clearly nonsense. The problem

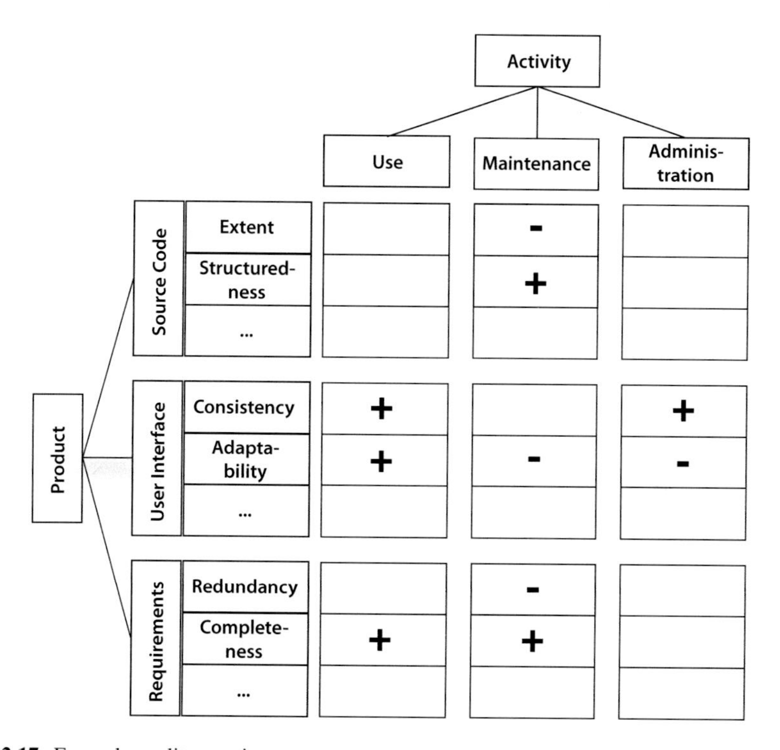


**Fig. 2.17** Example quality matrix

lies in the incompleteness of the depicted model that does not include tools like integrated development environments.

As we mentioned above, we often omit explicit properties for activities where it is clear that we mean the effectiveness and efficiency of the activities. Nevertheless, in some areas of the quality model, it is useful to have the possibility to be more precise. For example, when we describe usability aspects, we specify properties of the usage activity such as *satisfaction* or *safety* [217]. Then we can analyse the different aspects of the usage of the system more accurately. In many cases, however, it is more interesting to find a suitable decomposition of the relevant activities and only consider effectiveness and efficiency of those.

## 2.4.8 Tool Support

Comprehensive quality models contain several hundred model elements. For example, the maintainability model that was developed for a commercial project in the field of telecommunication [27] has a total of 413 model elements consisting of 160 product factors (142 entities and 16 properties), 27 activities and 226 impacts. Hence, quality models demand a rich tool set for their efficient creation, management and application just like other large models, e.g. UML class diagrams.

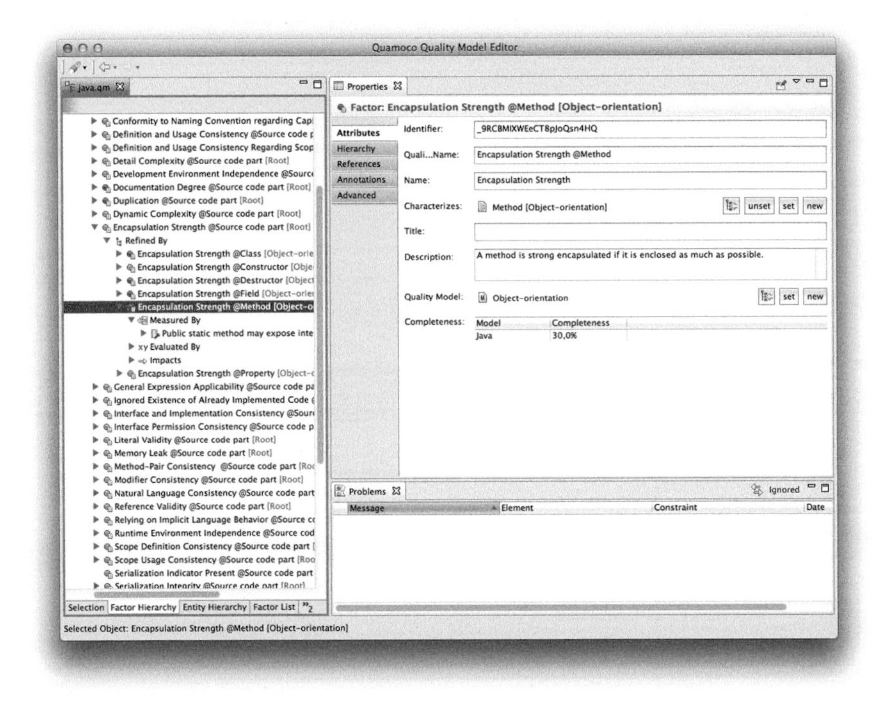


**Fig. 2.18** The Quamoco quality model editor

Because our quality models are based on an explicit meta-model, we are able to provide a model editor that does not only allow the initial development of quality models but also supports other common tasks like browsing, persistence, versioning and refactoring (Fig. 2.18). In Quamoco, we have developed a tool to define this kind of large and detailed quality models. Besides the easier creation and modification of the model, this has also the advantage that we can automate certain quality assurance tasks. For example, by using the tool we can automatically generate customised review guidelines for specific views. In addition, we have implemented a direct coupling with the quality dashboard tool ConQAT[10] so that the results of diverse quality analysis tools, tests and reviews can be read and aggregated with respect to the quality model.

One of the most powerful features of the model editor is the automatic generation of guideline documents from the quality model. This enables us to transfer the abstract definition of quality stored in the model to a format developers are familiar with. Moreover, unlike classic, handwritten guidelines, the automatically generated ones are guaranteed to be synchronised with the quality model that explicitly

[10]http://www.conqat.org/.

captures the understanding of quality within a project or a company. You can tailor guideline documents for specific needs by defining selected views on the model. For example, a guideline document could be specifically generated for documentation review sessions.

### 2.4.9 Summary

The Quamoco approach to modelling quality advances on previous approaches to model quality in the following issues:

#### Flexibility in the Quality Aspects

The Quamoco quality modelling approach allows to completely include the standard ISO/IEC 25010 quality models. We can use product quality characteristics or quality in use activities as quality aspects. If we need other views on the product factors, it would be no problem to include other quality aspect hierarchies.

#### Unambiguous Decomposition Criteria

Previous hierarchical models often exhibit a "somewhat arbitrary selection of characteristics and sub-characteristics" [121,122]. We claim this is because previous models lack a clearly defined decomposition criterion. For example, it is not clear how *self-descriptiveness* relates to *testability* in Boehm's quality characteristics tree (Fig. 2.16) if one is not satisfied with a trivial "has something to do with". The activity-based approach overcomes this shortcoming by separating aspects that are often intermingled: activities, entities and properties. This separation creates separate hierarchies with clear decomposition criteria.

#### Explicit Meta-Model

Unlike most other approaches known to us, our approach is based on an explicitly defined meta-model. This enables us to provide a rich set of tools for editing and maintaining quality models. Most important, the meta-model is a key for the model-based quality control approach outlined before. Additionally, the meta-model fosters the conciseness, consistency and completeness of quality models as it forces the model builders to stick to an established framework and supports them in finding omissions.

Complete Model

The most important advantage of Quamoco quality models is the complete chain from abstract quality aspects over product factors to concrete measurements. This allows us to define comprehensible evaluation methods as well as to refine quality requirements to be concrete and measurable.

## 2.5 Quality Model Maintenance

The quality model that defines the quality requirements of a system is, like the system itself, not a static entity. On the contrary, it needs to be updated and adapted over time. I will describe sources for model changes and give methodological guidelines for model maintenance.

### *2.5.1 Sources for Model Changes*

As in software, changes of the model can come from various sources. Most often, changes in the used technology, programming language or framework induce changes in the model. The providers of these technologies introduce new concepts, new language constructs or new libraries which all have their own product factors that we need to cover in the quality model.

In addition, the model is not fault free. It is built by humans and therefore there are incorrect model elements or relationships between model elements. For example, there might be an impact from the conciseness of identifiers in the code to the understanding of the end user. This is a mistake of the modeller who mixed up the activities. We need to correct this kind of faults. In the same example, the modeller probably then forgot to include an impact to the understanding of the developer. We also need to correct these omissions.

A different kind of omission is, if we find new product factors. Empirical research aims at finding such factors that significantly influence different quality factors. A new study might find that the level of redundancy in the source code has an influence on the number of faults in the system and, thereby, it has an impact on how often an end user experiences failures. We need to change the model to include such new factors. Similarly, empirical research or experience in a company can improve the knowledge about a product factor, e.g. what is the important property that influences quality or how strong is the impact. These refined informations also lead to changes in the model.

Finally, you can also change the quality goals and thereby induce a change in the model. If you realise that your system has become security-critical over time, you can create corresponding quality goals. To fulfil these goals, we need to extend the quality model with product factors that impact the quality aspects related to these goals. Analogously, quality goals can decrease in importance and be a reason to remove certain quality factors from the model.

### 2.5.2 Analysis and Implementation

A model change is similar to a software product change. First, we need to analyse the change request and then implement it in the model. The analysis determines what is wrong or missing in the model, which elements of the quality model are affected. An entity, a property or a whole product factor can be affected. For example, its name or description might be misleading. The relationships between factors can be wrong or missing. For example, an impact between two factors might be missing.

After the analysis, change the model either by changing element names, descriptions or additional information, by adding new elements or by deleting existing elements. It is important that you check the consistency of the model after each change. Removing model elements might affect other parts of the model that have then impacts for which the target is missing, for example. A good model editor can help you in showing you warnings for actual or potential inconsistencies.

### 2.5.3 Test

After you performed all changes and checked that the model is consistent again, you need to test if your model still works correctly in your quality control loop. This is important, because a quality model consists of a lot of text that carries semantics. Consistency and correctness of these texts cannot be checked automatically. Also evaluations that you specified can only be checked syntactically. If they make sense is still open. For that, we need to perform tests in the control loop.

It is useful to have a standard test system that you evaluate with the quality model after you performed changes on the model. The test system should not be different before and after the model change. If you keep the evaluation results of each test run, you can compare the results. There should be differences, because you changed the model. The differences, however, should only be the ones you expected by changing the model. If any unexpected differences occur, check if they are reasonable for the test system or if you introduced new faults into the model. Problems that you detect here become new change requests for the model and lead to re-analysis and re-modelling.

In case you worked on descriptions of model elements that you also use to generate review checklists and guidelines, a useful test is also to give it to several developers and testers and ask them for feedback. They are the ones who need to understand what you wanted to say with the description. Make sure they really understand.

### 2.5.4 *Checklist*

- Do you regularly check relevant publications for new product factors or improved knowledge about product factors?
- Do you feed back problems with the model that you found in the quality control loop into model changes?
- Do you regularly reevaluate your quality goals and change your quality model accordingly?
- Before you perform changes, have you analysed which model elements are affected?
- After you had performed a change, have you checked the model's consistency?
- Do you have a test system for which you keep the evaluation results?
- For each model change, do you perform test runs on test systems?
- For changes in checklists and guidelines, do you ask developers and testers for feedback?

## 2.6 Detailed Examples

We look in this section into three detailed examples of quality models to provide an illustration of the core concepts. The first example quality model follows directly the Quamoco approach. It is the performance efficiency part of the Quamoco base model. The second example uses the Quamoco concepts but only to describe quantifiable quality requirements without evaluation specifications. In particular, it uses an activity-based decomposition of attacks to describe security. The third example contrasts the other two, as it is a reliability growth model that illustrates a very focused prediction model.

### 2.6.1 *Performance Efficiency Part of the Quamoco Base Model*

The Quamoco base model is one of the main results of the Quamoco project. It captures the experiences and discussions of all experts who took part in the project. The whole base model is too large to describe here, but it is available on the Quamoco web site.[11] Our aim here is to illustrate the quality model concepts, and therefore, we concentrate on one smaller part of the model: the performance efficiency model with measurements for Java.

Performance efficiency is the quality characteristic of ISO/IEC 25010 describing how well the software uses the resources for a fast execution. In the standard, it is

[11] http://www.quamoco.de.

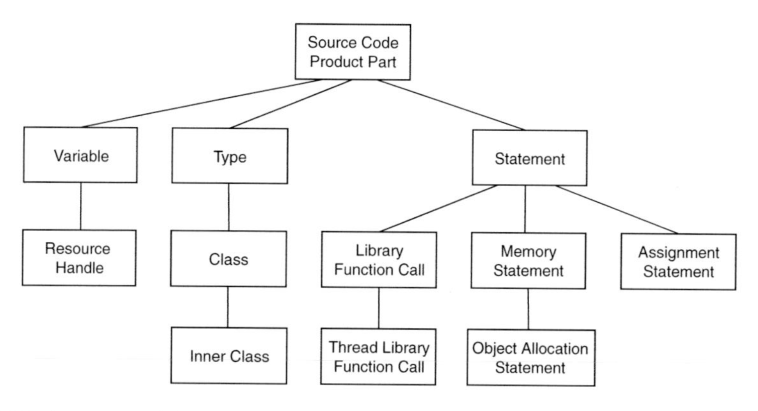


**Fig. 2.19** An excerpt of the entity tree for the performance model

defined as "The performance relative to the amount of resources used under stated conditions" [97]. It is decomposed into two sub-characteristics: resource utilisation and time behaviour. Resource utilisation is "the amounts and types of resources used when the product performs its function under stated conditions in relation to an established benchmark", and time behaviour is "the response and processing times and throughput rates when performing its function, under stated conditions in relation to an established benchmark" [97]. We have not defined a benchmark in Quamoco but looked at statically visible problems in the source code that can impact these quality characteristics.

Figure 2.19 shows an excerpt of the entities tree of entities with an impact on performance efficiency or one of its sub-characteristics. You can see that various common source code parts have an influence, such as assignment statement, but several entities have special purposes that already suggest a relation to performance efficiency, such as resource handle or object allocation statement.

Using these and other entities, there are more than 50 product factors with measurements for Java alone influencing performance efficiency and its sub-characteristics. We will look at several examples that have a relatively strong impact according to our calibration of the base model.

The strongest impact on resource utilisation has the product factor describing resource handles: [Resource Handle | DEFINITION AND USAGE CONSISTENCY]. In the model, it is defined that "A resource handle's definition and usage are consistent if its definition allows usage in the intended way and if it is used in the defined way". The resource handles in Java are, for example, streams or databases. The corresponding measures for this product factor hence are static checks about closing and cleaning up streams or databases. FindBugs provides some rules to check for these kinds of bug patterns. We weighted all of them the same but use different linear decreasing functions depending on the number of warnings for each FindBugs rule.

A second example of a product factor is [Inner Class | UNNEEDED RESOURCE OVERHEAD]. Inner classes are useful for various purposes, and they normally keep in Java an embedded reference to the object which created it. If this is not used, we could reduce the size of the inner class by making it static. These and similar related problems can create an unneeded resource overhead for inner classes, which is expressed by this product factor. Again, we use a set of FindBugs rules to measure the size of the problem for a given software.

A third example is [Object Allocation Statement | USELESSNESS] which we will use to discuss an evaluation in detail (see also Sect. 4.2). We defined this product factor in the quality model with "An object allocation statement is useless if the object is never used". For Java, this is measured using two measures which use FindBugs intstruments: *Exception created and dropped rather than thrown* and *Needless instantiation of class that only supplies static methods*. Each measure has a weight of 50 %. Both are modelled with linear increasing functions. The exception measure adds its 50 % if the density of exceptions created and dropped is above 0.00003 and the class instantiation measure if there is any such warning. Together they describe on a range from 0 to 1 to what degree there are useless object allocation statements in the software. This product factor influences resource utilisation and time behaviour (among others). As mentioned before, resource utilisation has more than 50 product factors influencing it of which [Object Allocation Statement | USELESSNESS] has a weight of 6.65 %. Together with the other product factors, this will give an evaluation of the quality aspect. You will find more details on the quality evaluation method in Sect. 4.2.

In summary, this model can only partially evaluate the inherently dynamic quality "performance efficiency". Nevertheless, we found over 50 product factors with several measures for each product factor that can be determined statically. This is an interesting aspect because it is far cheaper and easier than dynamically testing based on a benchmark. For exact performance efficiency analyses, however, we will need to add these dynamic analyses. For an early and frequent look onto this quality factor, however, this part of the Quamoco base model is suitable. In the spirit of COQUAMO, you can see it as a quality indicator during development.

### 2.6.2 Web Security Model

Malicious attacks on software systems are a topic with high public visibility as average users can be affected. The level of vulnerabilities is still high today. For example, the CERT[12] reported 6,058 total vulnerabilities for the first 9 months of 2008. These attacks have a strong financial impact. In an E-Crime Watch Survey,[13] it is stated that on average for each company security attacks result in a monetary loss of $456,700 in 12 months.

Therefore, software security is still a large and important problem in practice. It affects not only financial aspects but also ethical issues such as privacy. Hence, it is

---

[12] http://www.cert.org/stats/vulnerability_remediation.html.

[13] http://www.csoonline.com/documents/pdfs/e-crime_release_091107.pdf.

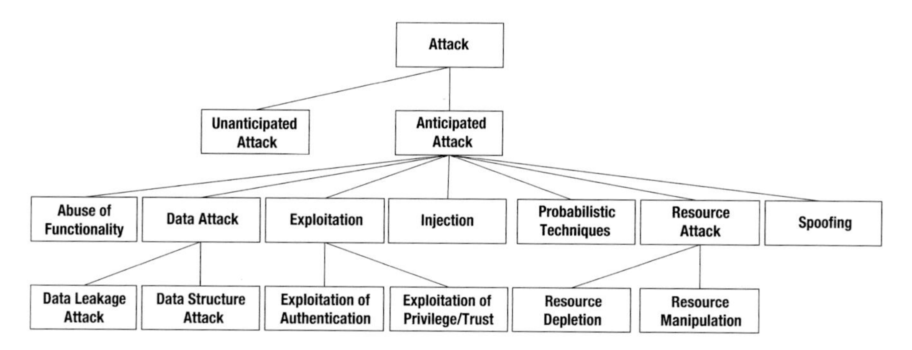


**Fig. 2.20** Excerpt of the upper layers of the attack sub-tree of the activity hierarchy [214]

an important goal in software development to produce *secure* systems. This especially holds for web systems that are usually accessible in public networks. Secure web systems begin with the specification of security requirements. To become effective, they need to be clear and precise so that they are a real guidance for their implementation. This way, vulnerabilities can be prevented or at least reduced.

Despite the importance of high-quality security requirements for web systems, in practice they are often not well documented or not documented at all. This results also in often only poorly tested security properties of the final systems.

For specifying and verifying web security requirements, we created a web security instance of an activity-based quality model. Most important in this instance is to add attacks to the activity hierarchy. These are activities that need to be negatively influenced. First, we have to differentiate between anticipated attacks and unanticipated attacks. A major problem in software security is that it is impossible to know all attacks that the system will be exposed to, because new attacks are developed every day. Hence, to assure quality, we need to employ two strategies: (1) prepare the system against anticipated attacks and (2) harden the system in general to avoid other vulnerabilities. For the classification of the attacks, there are several possible sources. We rely on the open community effort *Common Attack Pattern Enumeration and Classification* (CAPEC) [83] that is led by the U.S. Department of Homeland Security. In the CAPEC, existing attack patterns are collected and classified. The attacks are organised in a hierarchy that is adapted for the activities hierarchy (Fig. 2.20).

The creation of the product factors is far more complicated. The product factors need to contain the available knowledge about characteristics of the system, its environment and the organisation that influence the described attacks. We employed various sources for collecting this knowledge including the ISO/IEC 27001 [106], the Web Guidelines,[14] OWASP [216] and the Sun Secure Coding Guidelines for the Java Programming Language [194]. However, two main sources were used

[14]http://www.webguidelines.nl/.

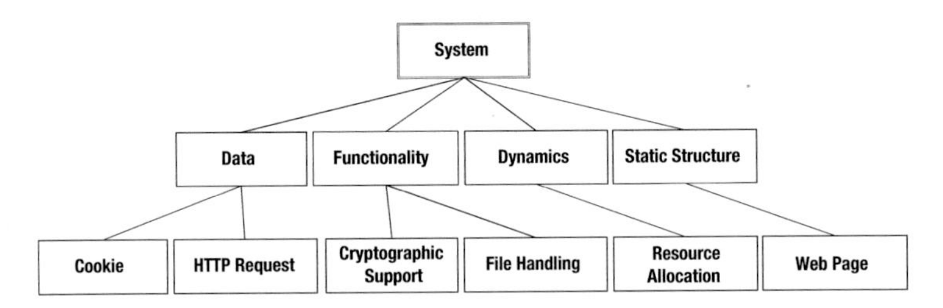


**Fig. 2.21** Example entries of the system sub-tree from the entities [214]

because they constitute large community efforts and hence provide consolidated knowledge: specific parts of the *Common Criteria* (CC) [43] and the *Common Weakness Enumeration* (CWE) [84]. The Common Criteria describe requirements on a system that should ensure security with a focus on what the system "shall do". The CWE looks at security from the other direction and describes reoccurring weaknesses in software systems that lead to vulnerabilities that are exploited by attacks. Therefore, these two sources combined give a strong basis for the product factors.

We cannot describe the incorporation of the sources in all details in this example, but we give some instances on how knowledge from the sources has been modelled in our security model. For this, we use a sub-tree of the entity hierarchy for the system as depicted in Fig. 2.21. The system consists of Data and Functionality. Furthermore, it has Dynamics and a Static Structure. These entities have then again children. For example, data can be a Cookie or an HTTP Request. Interesting functionality can be Cryptographic Support or File Handling.

Many of the entries in the quality model that have their origin in the Common Criteria are modelled as a part of Functionality because they mainly describe behavioural aspects that nevertheless are important for security. An example that is shown in Fig. 2.21 is the cryptographic support of the system. Following the CC, this can be decomposed into Cryptographic Key Management and Cryptographic Operation. A further part of Cryptographic Key Management is the Cryptographic Key Generation. The CC defines a requirement for that key generation that it shall be in accordance with a specified algorithm and specified key sizes. In the model, we express that by using the property APPROPRIATENESS for Cryptographic Key Generation. The resulting factor [Cryptographic Key Generation | APPROPRIATENESS] is textually described by "The system generates cryptographic keys in accordance with a specified cryptographic key generation algorithm and specified cryptographic key sizes that meet a specified list of standards". Unfortunately, the CC does not contain any description of impacts. This would make the standard more useful because the motivation to use these requirements would be higher. Hence, we complete the information using other sources. In this case, the CAPEC contains possible solutions and mitigations in the description of the *cryptanalysis* attack that includes the recommendation to use proven cryptographic algorithms with recommended key sizes. Therefore, we include the corresponding negative impact of [Cryptographic Key Generation | APPROPRIATENESS] on Cryptanalysis.

In contrast to the CC, the *Common Weakness Enumeration* mainly provides characteristics of the system and especially the kind of code that should be avoided. We include these characteristics into the model in this negative way with a positive influence on the attacks, i.e. making attacks easier. Another possibility is to reformulate the weaknesses as strength that are needed with negative influence on attacks. We used both possibilities depending on which option was more straightforward to model.

Several weaknesses in the CWE are not aiming at specific attacks but describe characteristics that are indicators for possible vulnerabilities. We model these as factors that have an impact on unanticipated attacks. An example from the CWE that is in our security model is *dead code*. Several parts of the code can be superfluous such as variables, methods or complete classes. For a variable, we can model that as a positive impact of [Variable | SUPERFLUOUSNESS] on Unanticipated Attack.

### 2.6.3 Reliability Model

Reliability theory and models for estimating and predicting reliability have been developed for several decades. Reliability models are a very different type of quality model than the models we build with the Quamoco approach. They have a certain usage in practice and their application areas. We explain some basics and summarise the merits and limitations in the following. The general idea behind most reliability models is that we want to predict the future failure behaviour – and thereby the reliability – of a software based on actual failure data. Reliability is already defined as probability. That means that we use data from failures as sample data in a stochastic model to estimate the model parameters. There are other kinds of reliability models that use different mechanisms, but they are not as broadly used. The failure data that comprises the sample data can be one of two types: (1) time between failure (TBF) data or (2) grouped data. The first type contains each failure and the time that has passed since the last failure. The second type has the length of a test or operation interval and the number of failures that occurred in that interval. The latter type is not as accurate as the first, but the data is easier to collect in real-world projects.

Having used the sample data to estimate the model parameters, we can use the model with these parameters to predict future behaviour. That means the model can calculate useful quantities of the software at a point in time in the future. Apart from reliability the mostly used quantities are the expected number of failures up to a certain time $t$ (often denoted by $\mu(t)$) and its derivative, the failure intensity (denoted by $\lambda(t)$). The latter can intuitively be seen as the average number of failures that occur in a time interval at that time $t$. Based on these quantities we can also predict, for example, for how long we have to continue testing to reach a certain failure intensity objective.

An excellent introduction to the topic is a book by Musa [158]. A wide variety of partly practical relevance is described in a handbook [139]. Finally, the most detailed description of these models and the theory behind them can be found in [159].

#### Execution Time and Calendar Time

Experience indicates that the best measure of time is the actual CPU execution time [159]. The reliability of software as well as hardware that is not executed does not change. Only when it is run there is a possibility of failure and only then there can be a change in reliability. The main cause of failure for hardware is considered the wear-out. Therefore, it is possible to relate the execution time to calendar time in some cases. For software this seems to be more difficult and hence execution time should be used.

However, CPU time may not be available, and it is possible to reformulate the measurements and reliability models in terms of other exposure metrics: clock time, in-service time (usually a sum of clock times due to many software applications running simultaneously on various single- or multiple-CPU systems), logical time (such as number of executed test cases, database queries or telephone calls) or structural coverage (such as achieved statement or branch coverage). One hundred months of in-service time may be associated with 50 months (clock time) of two systems or 1 month (clock time) of 100 systems. All of these approximations are also referred to as *usage time*.

In any case, some combination of statistical sampling with estimates of units sold is much better than using calendar time because of the so-called loading, or ramping, effect [111]. If this effect is not accounted for, most models assume a constant usage over time which is not reasonable under many practical circumstances. The difficulty is to handle differing amounts of installations and testing effort during the life cycle of the software.

#### Parameter Estimation

Parameter estimation is the most important part of applying a software reliability model to a real development project. The model itself is hopefully a faithful abstraction of the failure process, but only the proper estimation of the model parameters can fit the model to the current problem. There are several ways to accomplish that, but a consensus seems to be reached that a Maximum Likelihood approach on the observed failure data fits best to most models. Another possibility is to use other software metrics or failure data from old projects as basis for the estimation, but this is only advisable when no failure data is available, i.e. before system testing.

### Merits and Limitations

The usage of stochastic reliability models for software is a proven practice and has been under research for decades now. Those models have a sound mathematical basis and are able to yield useful metrics that help in determining the current reliability level and in deciding the release problem.

However, the models are often difficult to use because (1) some understanding of the underlying statistics and (2) laborious, detailed metrics documentation and collection are necessary. Moreover, the available tool support is still not satisfying. The main deficiency is, however, that the models are only applicable during system test and field usage and even during system test they depend on usage-based testing, i.e. they assume that the tests follow the operational profile that mirrors the future usage in the field. The approach to use other metrics or failure data from old projects is very fragile. Especially to use other software metrics has not led to a satisfying predictive validity. This narrows the merit of those models.

# Chapter 3
# Quality Planning

Quality is not a fixed property of a software system, but it depends on the needs and goals of the stakeholders. Therefore, we have to carefully plan what quality the system should have. This involves that we identify the stakeholders and understand their needs and map those needs to technical properties and, finally, quality requirements of the system that the developers will implement. In addition, we need to plan not only *what* quality we want to build but also *how* we will build and assure it.

## 3.1 Model Building and Quality Requirements

Quality requirements are usually seen as part of the *non-functional* requirements of a system. Those non-functional requirements describe properties of the system that are not its primary functionality. "Think of these properties as the characteristics or qualities that make the product attractive, or usable, or fast, or reliable" [183]. Although this notion of non-functional requirements is sometimes disputed, there always exist requirements that relate to specific qualities of the system [76]. We call those demands *quality requirements*. We exclude process-related quality requirements here, because we concentrate on product quality.

Quality requirements are an often neglected issue in the requirements engineering of software systems. A main reason is that those requirements are difficult to express in a measurable way what also makes them difficult to analyse [167]. One reason lies in the fact that quality itself "[...] is a complex and multifaceted concept" [71]. It is difficult to assess and thereby also specifying quality requirements is a complex task. In particular, incorporating the various aspects of all the stakeholders is often troublesome. Hence, the problem is how to elicit and assess quality requirements in a structured and comprehensive way.

S. Wagner, *Software Product Quality Control*, DOI 10.1007/978-3-642-38571-1_3,

The main approaches to elicit quality requirements are:

- Checking different requirements types and building prototypes [183]
- Using positive and/or negative scenarios (use cases and misuse cases) [3]
- Refining quality goals by goal graphs [45]

An example for checking requirements types is the checklists for Service Level Agreements (SLA) used in ITIL.[1] In use cases, quality requirements are attached as additional information such as the maximal response time of the system in that use case. Goal graphs often break down quality goals to logic statements to at last operationalise them as functional requirements. Although I believe that all of the three approaches are valid, important and best used in combination, the incorporation of modern quality models (Chap. 2) into the requirements engineering process can improve the resulting specification by defining more structure and precision in measurements. Moreover, we reuse quality models from one project to the next and hence a common understanding of quality can build up between the stakeholders, for example, about terminology, interrelations of quality aspects or implications to costs. In this way, specifying quality requirements and building quality models fuse into one process. While we always add new information to a quality model when building it, we specify specific requirements and choose the suitable parts of existing quality models in the requirements engineering process.

For a structured requirements engineering approach using a quality model, we suggest to go from the stakeholders and their general goals to quality goals and to refine these with quality aspects and product factors to quality requirements. If there is an existing Quamoco quality model, we use the structure induced by it to elicit and refine the quality requirements. The elicitation and refinement are strongly oriented at using the two main hierarchies contained in the quality model and aim at refining the requirements to quantitative values as far as possible. We specifically make use of the idea of activity-based quality models (see Sect. 2.4) because they directly support the derivation from quality goals from stakeholders. It is straightforward to identify activities of stakeholders and elicit how we should support these activities.

This approach depends on the availability of a suitable quality model. Ideally, an appropriate quality model exists that shows the needed quality aspects and entities. This will often not be the case, however. Then the same approach also supports to build and extend the existing quality model or the Quamoco base model in parallel to specifying the requirements. Useful quality models are detailed and therefore at least some parts are specific. Hence, any organisation faces the challenge of building appropriate quality models that cover its goals and technologies. Yet, the long-term objective is to come to a quality model that is stable over most projects of your organisation, which means that we save effort by reusing most of the model. Our experience shows that the reuse ratio over projects increases over time

[1] IT Infrastructure Library: http://www.itil-officialsite.com.

1. Identify relevant stakeholders.
2. Define general goals.
3. Analyse relevant documents and elicit new information.
4. Choose and define activities.
5. Define quality goals.
6. Identify affected artefacts and choose entities.
7. Choose entities and analyse relevant material.
8. Choose factors and define new product properties.
9. Specify quality requirements.

**Fig. 3.1** The model building and requirements engineering steps at a glance

(see Sect. 5.3). We provide a step-by-step approach to specify quality requirements and build a quality model starting from general goals over quality goals down to concrete quality requirements. We further illustrate the approach by two examples and finally give condensed experience in the form of a checklist.

### *3.1.1 Step by Step*

The process of specifying quality requirements and building an integrated quality model can be complex and elaborate. To help you structure this process, I provide a step-by-step approach that will guide you along the way. You can find an overview of the steps in Fig. 3.1. For illustration, we use the running example of a web shop for books.

#### Step 1: Identifying Relevant Stakeholders

The first step is, similar as in other requirements elicitation approaches, identifying the stakeholders of the software system. For quality requirements, this usually includes users, developers and maintainers, operators and user trainers. Depending on your context, other stakeholders can also be relevant [178]. The stakeholders are central to our approach, because how you have to instantiate the abstract concept "quality" depends on what they need.

Consider, as an example, a web shop. It has at least two types of users: customers and sellers. Further stakeholders are developers and maintainers as well as administrators. In addition, we consider a marketing manager who is responsible for the commercial success of the web shop. All these stakeholders have needs that the system should satisfy, and those needs define quality on the most abstract level. In other words, we take the *user view* on software quality (see Sect. 1.3).

Step 2: Define General Goals

Next, to get a grasp on the diffuse stakeholder needs, we need to define general goals that stakeholders have for the software system. These goals will shape what we are going to include in the quality requirements and the quality model. Furthermore, goals give a rationale for the requirements we will specify and help us in understanding their relevance. In many cases, you can find goals in marketing plans or documentations. The goals defined should especially include goals of the users, e.g. what activities they want supported, and of the development organisation, e.g. the number of customers they want to achieve. In the web shop example, the marketing manager could set the goal to reach a market share of 20 % in Germany.

In the discussion about quality models and measures, we also mentioned GQM (Chap. 2) as an approach to structure a measurement programme. The GQM approach suggests to start from measurement goals and to break these down into questions we want to answer to reach the goal. For each of these questions, we then define metrics as (partial) answers to the questions.
Our quality planning approach from this chapter is similar in the structure and basic idea. We want to be driven by the goals of the stakeholders to define focused measurements necessary to assess the goals. GQM is, however, more general as you can have goals that do not relate to quality goals and quality requirements. Nevertheless, you can understand our approach in conjunction with quality models as GQM patterns that you reuse over projects.

**Sidebar 1:** Goal-Question-Metric

Step 3: Analyse Relevant Documents and Elicit New Information

The general goals defined in step 2 need to be refined in this step. Therefore, we search for existing documents that help in this refinement. This usually includes process documentation (development process and business processes), user documentation, usage scenarios and standards. In addition, you can use any requirements elicitation methods, such as interviews or focus groups, to get new information, especially new and relevant usage scenarios. For the web shop, we elicit usage scenarios and analyse to what extent they support our general goal. We develop a set of business processes and use cases and usage scenarios that show how customers and sellers trade goods on the web shop as well as how the administrator and maintainer make changes to it. Our analysis in this case shows that these scenarios reflect the user needs enough so that we could reach the set market share.

Step 4: Choose and Define Activities

After you have identified the stakeholders with their goals and usage scenarios, an existing quality model can help you to derive the activities they perform on and with the system. For example, the activities for the maintainer include *concept location*, *impact analysis*, *coding* or *modification* [51]. Especially the activities of the user can be further detailed by additional usage scenarios. In usability engineering, they tend to talk about *tasks* which means the same as what we call *activities*. The activities also give us a connection to the *value-based view* on product quality (Sect. 1.3), because they incur costs and create benefits.

Step 4 is the first step in which you build a concrete part of the quality model if something is missing for your context. If you use a quality in use or activity-based quality models, it needs a hierarchy of activities that are relevant for quality as building blocks. If you rely on product quality characteristics, just describe the activities separately. We will transform the activities into quality characteristics later. We have to check if there are existing activities and if they cover the general goals of the stakeholders. We might derive new activities from the general goals and documents that we analysed in the previous step. In the development process documentation, we find the relevant activities in development and maintenance, in the business process documentation the activities that the users need to perform. Usage scenarios give in detail the steps that users perform. For the web shop, we define activities such as *searching for books*, *putting books in the basket* or *paying*. These are the main activities that are important for our general goal. To form a hierarchy, we set *usage* as the parent quality aspect.

#### Step 5: Define Quality Goals

On the level of the activities of step 4, we define quality goals. These quality goals comprise how well we want to support each quality aspect. If we plan to extend and maintain our system, we probably want to support all maintenance activities that should be described in our development process documentation. We defined important activities for the web shop, e.g. searching for books. For reaching our general goal of 20 % market share, we need to support as many user groups as possible and reduce barriers for using the web shop. People with disabilities have special requirements on how they perform these activities. Hence, the quality goal is that the defined activities can be performed by people with disabilities.

We also rank the activities of the relevant stakeholders according to their importance. This results in a list of all activities of the relevant stakeholders. On top of this list are the most important activities, the least important at the bottom. Importance hereby means the activities that you expect to be performed most often and which are most elaborate. The justification can be given by expert opinion or experiences from similar projects. We will use this list in the following to focus the definition and refinement of the requirements:

Now, we need to answer the question how well we want the activities to be supported. The answers are in essence qualitative quality goals for the software system. There are various methods for evaluating and prioritising the activities and hence the quality goals. We use an elementary method that assigns qualitative statements to activities. For example, if we expect rather complex and difficult concepts in the software because the problem domain already contains many concepts, the activity *concept location* is desired to be *simple*. These qualitative

*Quality Function Deployment* (QFD) is a technique for quality planning in the sense that it is a management task which has the aim to find out what quality the customer wants and how we can reach that quality with our product. QFD has its origins in the area of manufacturing industries, but it has been used for software as well.
The front end of QFD is a specific requirements engineering process. We gather the *voice of the customer* (VOC) by common means like interviews or questionnaires. We then analyse the VOC by grouping and merging different voices. We do that together with customers in workshops or by building prototypes and user tests. Finally, we begin to work on the *House of Quality* (HOQ) to add technical realisations to the customer needs. Here is a small example of an HOQ for a software system:

| | | GUI | | Progr. Lang. | |
|---|---|---|---|---|---|
| | | Native | Swing | Java | C |
| Easy to use | Quick learning | ✔ | ✖ | | |
| | Intuitive control | ✔ | ✖ | | |
| Reli-able | No total crashes | | | ✔ | ✖ |
| | Seldom malfunct. | | | ✔ | ✖ |

At the left, we have some example customer needs for a software system. The columns are possible technical realisations. The check marks and crosses show whether a technical realisation is important for the need or if it is even obstructive. The roof of the house shows similar relationships between the technical realisations. For example, a Swing GUI works well with Java but does not work with C. With this analysis, we prioritise and resolve trade-offs. Then we can start a second house on a lower level of detail that shows design decisions specific for the system's software architecture.
The general idea behind QDF is similar to our quality planning approach described in this chapter. Even the HOQ resembles some similarities with a quality model. On the left side, there are quality goals and we describe relationships to technological factors. QFD adds to our approach that it handles functional requirements similarly to quality requirements and it makes influences between technologies and product factors explicit in the roof of the HOQ. So we can map our approach into QFD. We give, however, more concrete steps that lead you from general goals to measurable quality requirements.

**Sidebar 2:** Quality Function Deployment

statements are needed for all the activities. Depending on the amount of activities to be considered, the ones at the bottom of the list might be ignored and simply judged with *don't care*.

At this point, we can transform the quality goals based on activities into product quality attributes, if this is our preferred way of modelling quality. For several high-level activities, there is a direct counterpart in the quality characteristics of

**Table 3.1** Mapping of activities to quality factors

| Activity | Quality factors |
|---|---|
| Effective usage | Functional suitability, reliability |
| Efficient usage | Usability, reliability |
| Usage with specific hardware | Performance efficiency |
| Usage with other software | Compatibility |
| Usage without risk | Security |
| Attack | Security |
| Maintenance | Maintainability |
| Porting | Portability |

ISO/IEC 25010. For example, effective and efficient maintenance can be expressed by *maintainability* or the effective usage by people with disabilities can be expressed by *accessibility*. An example list of this transformation is given in Table 3.1. Such a direct transformation is not always possible. Specific activities, such as *concept location*, need to be mapped to a more abstract quality attribute, such as *analysability*.

### Step 6: Identify Affected Artefacts and Choose Entities

Quamoco quality models do not only specify activities and quality goals, but they make the relation to product factors explicit. Therefore, we need to find the parts of the product, i.e. all technologies and artefacts, that have an influence on the defined activities. The problem often arises that almost the complete product has some kind of influence on the activities of interest. It is important to concentrate here on the most relevant ones. Otherwise, the quality model and the building effort might become too large to be manageable.

Our web shop is a rich internet application. Hence, web technology such as HTML and Java Script is relevant. To make the web shop accessible affects mostly the user interface. For that reason, the usage scenarios and the HTML output pages are the main artefacts we need to look at.

In the quality model, we describe artefacts and technologies with entities. In an existing quality model, the entities hierarchy contains the entities of the software and its environment that are relevant for the quality of the system. The entities hierarchy organises them in a well-defined, hierarchical manner that fits to the task of refining the quality goals elicited based on the activities. The quality model itself is a valuable help for this. It captures the influences (of properties) of entities on the activities or quality characteristics. Hence, we only need to follow the impacts the other way round to find the entities that have an influence on a specific quality aspect. If these influences are incomplete in the current model, this step can also be used to improve it. This way, consistency with the later quality assurance is significantly easier to achieve. For specifying the refined quality requirements, the product factors related to the entities can be used. For example, a detailed requirement might be that each object must have a state accessible from outside, because this has a positive effect on the *test* activity.

### Step 7: Analyse Relevant Material for Artefacts

In step 7, we need to find the basis for properties of the identified artefacts to form product factors that have an impact on the identified activities. For this, we search for relevant material from all sources we can find. Usually, there are existing checklists and guidelines used internally in the company and from external sources. For example, Sun released a coding guideline for Java in which we can look for product factors from Java source code. Several standardisation bodies produce relevant standards for software quality, e.g. ISO, IETF, IEEE and W3C. Although the information in standards tends to be abstract, there are often parts that can be easily integrated into a quality model.

For areas that are not well covered by existing materials, it is also advisable to interview experts inside the development organisation, e.g. developers, usability experts or IT administrators. For specific properties, there also might be an empirical study that investigated their impacts. Nowadays, many scientists make their papers available at their home pages. A good source for freely accessible scientific papers is Google Scholar.[2] The main source of information that we use for the accessibility of the web shop comes from the W3C that developed the working draft *WAI-ARIA Authoring Practices* for accessible rich Internet applications.[3]

### Step 8: Choose Factors and Define Product Properties

In this step, we finally transcend from the *user view* to the *product view* on product quality (Sect. 1.3). We come to this view by defining concrete product factors. For that, we compare the analysis from the relevant material with the product factors that are already contained in our quality model. If the model covers what the material gave us as information, this step is finished. Otherwise, we need to extend the model.

To this end, we use the analysed material to define product factors. The product factors contain entities together with their properties. The entities are the artefacts that we identified in step 5. The challenge is to use the appropriate level of abstraction and to organise them in an entity hierarchy. For Java source code, we have methods and attributes as parts of classes, which in turn are a part of packages. On the basis of the information from relevant studies, we then define the property *redundancy* of source code with a negative impact on the activities *source code reading* and *source code modification*.

---

[2]http://scholar.google.com.

[3]http://www.w3.org/TR/wai-aria/.

For the web shop, we find several properties in the *WAI-ARIA Authoring Practices*, which define how you should build dynamic web applications. For example, it states that for images that are used for buttons, alternative text has to be set. As our web shop uses image buttons, this property is relevant for the quality goal. We add the factor [Image Button | DESCRIPTIVENESS] to the quality model and amend it with impacts to the activity *Usage*, because all usages are affected.

Step 9: Specify Quality Requirements

Finally, we specify concrete quality requirements using defined product factors. This specification determines how we want this factor to be. For example, it is not realistic to require the source code to be completely free from redundancy, especially for a system that already exists. The effort would probably outweigh the costs. Therefore, a reasonable level of redundancy that can be tolerated needs to be specified. Hence, this also implies that we define measures for the factors in the model if they not already exist.

In our web shop, it has also to be decided whether *all* image buttons need to have alternative text. There might be special areas that are only used by internal administrators, who do not require this. To have these special areas accessible as well, we formulate the quality requirement, that all image buttons shall be descriptive. This requirement can then be checked in the concrete web shop to evaluate its quality.

The goal is to have quantitative and, hence, easily checkable requirements. Requirements on the product factor level should be quantitatively assessable (cf. [51]) even if with qualitative methods. For example, *needless code variables* can be counted and hence an upper limit can be given. Sometimes, it is easier to quantify requirements on the quality aspect level instead of the product factor level. On the quality aspect level, this would be, for example, that an average *modification* activity should take 4 person hours to complete. This can be used if a more detailed decomposition is not feasible for specific activity or quality attribute.

As a standard approach, we recommend to document in the requirements specification the complete trace from stakeholders and their general goals over activities and quality goals to product factors and concrete quality requirements. It gives you the most comprehensive picture of the needed quality together with rationales and measurements.

### *3.1.2 Example: Web Shop*

We summarise the web shop example to give you a more complete picture of the model building process and the kind of information you have to document. Nevertheless, we cannot discuss the whole model, but we picked several interesting parts. We summarise the example in two tables that show the information we collected: Table 3.2 represents steps 1–5, Table 3.3 steps 6–9.

**Table 3.2** Stakeholders with their goals, activities and derived quality goals for the web shop

| Stakeholder | Goal | Activity | Quality goal |
|---|---|---|---|
| User | Finding books easily | Searching for books | Simple and quick searching |
| Product manager | 20 % market share | Usage | Accessible for everyone |
| Developer | Quick changes | Modification | Simple modification |
| Administrator | Low hard disc needs | Hardware upgrade | Few upgrades |
| Attacker | Reveal credit card data | Attack | Secure paying |

**Table 3.3** Artefacts of the web shops with factors, impacted activities and requirements

| Artefact | Factor | Activity impact | Quality requirement |
|---|---|---|---|
| Image tag | Descriptive | Positive usage | All image tags shall have alternative text |
| Font size | Fixed | Negative usage | No font sizes shall be hard coded |
| Checkbox | Descriptive | Positive usage | All checkboxes shall have the appropriate ARIA role and states |
| Tooltip | Activatable | Positive usage | All tooltips shall be activated by mouse and keyboard |

We found that we have five relevant stakeholders for the web shop (Table 3.2): the users of the system, its product manager, its developers, the administrators responsible for running the web shop on their servers and potential attackers of the system. The user has the goal to find the wanted books easily. Hence, the activity is *searching for books* in the web shop. The quality goal is to support this search so that it is simple and quick. In the description of the 9-step approach, we already discussed that the product manager wants 20 % market share which reflects on the accessibility of the usage. Developers want to be able to make quick changes to the system without having to fix impacts all over it. Hence, the quality goal is a simple modification. Administrators aim at low hard disc consumption and therefore have the goal to perform few hardware upgrades.

Finally, potential attackers of the system will aim for the main assets, i.e. the credit card data from payments. Hence, for the attack activities, we define the goal that paying is secure. Note that for this stakeholder the derivation is different, because the derived quality goal is opposing to the general goal of the stakeholder. This is because an attacker is the only stakeholder with a malicious intent. For all other stakeholders, we assume that we want to support their general goal by quality goals.

In the following, we concentrate on the quality goal of the product manager that the web shop should be accessible for everyone. The exemplary artefacts we found important to analyse for ensuring this quality goal are image tags, font sizes, checkboxes and tooltips (Table 3.3). A basic accessibility requirement is that all image tags have alternative text so that readers can process them. This is modelled as the descriptiveness of image tags which has a positive impact on usage. Font sizes can be fixed or made variable with the help of CSS and JavaScript. For changing font sizes, we require that they are not hard coded. We make sure that checkboxes are descriptive by requiring that all checkboxes need to

**Table 3.4** Stakeholders with their goals, activities and derived quality goals for the instrument cluster

| Stakeholder | Goal | Activity | Quality goal |
|---|---|---|---|
| Driver | Travel to destination | Driving, TICS dialog | Correct, attractive TICS dialog, safe driving |
| OEM | Low costs | System integration, defect correction | Minimal defect correction, minimal hardware requirements |

**Table 3.5** Artefacts of the instrument cluster with factors, impacted activities and requirements

| Artefact | Factor | Activity impact | Quality requirement |
|---|---|---|---|
| Display position | Appropriate | Positive driving | The display tolerance amounts to ±1.5 degrees |
| Output data representation | Unambiguous | Positive processing | The engine control light must not be placed in the digital display with lots of other information |
| Output data representation | Adaptive | Positive perception | The output data representation must change with the driving situation |

have the appropriate ARIA roles and states. The ARIA standard defines a standard terminology for UI widgets. Each widget in a web application defines its role in this terminology so that alternative readers and input devices know how to handle the widget. For a checkbox, this is *checkbox*. The same case applies for states, which are for checkboxes *aria-checked="true"* or *aria-checked="false"*. Last, tooltips can be activated by hovering over it with the mouse, but for supporting other input devices, it needs to be activatable by the keyboard.

### 3.1.3 Example: Instrument Cluster

We discuss a second example, which is from the automotive domain, to further illustrate the quality requirements and model building approach. DaimlerChrysler published a sample system specification of an instrument cluster [30]. The instrument cluster is the system behind a vehicle's dashboard controlling the rev metre, the speedometer and indicator lights. The specification is strongly focused on functional requirements but also contains various high-level business requirements that consider quality aspects. The functional requirements are analysed in more detail in [127]. We mainly look at the software requirements but also at how the software influences the hardware requirements. Similarly to the web shop example, we summarise steps 1–5 in Table 3.4 and steps 6–9 in Table 3.5. The difference here is, however, that we assume that we have an existing quality model that fits for most parts.

We can identify two stakeholders for the quality requirements stated in [30]. The relevant requirements are mainly concerned with the user of the system, i.e. the *driver*. In general, drivers want to reach their destination. For that, they need to have a good view on all information, relevant information needs to be given directly, and their safety has to be ensured. To derive the corresponding activities, we can use the quality model for usability described in [217]. It contains a case study about ISO 15005 [89] which defines ergonomic principles for the design of transport information and control systems (TICS). The instrument cluster is one example of such systems. Hence, we can use the identified activities here. The distinction on the top level is in *driving* and *TICS dialog*. The former describes the quality aspect of controlling the car to navigate and manoeuvre it. Examples are steering, braking or accelerating. The latter means the actual use of a TICS system. It is divided into (1) *view*, (2) *perception*, (3) *processing* and (4) *input*. This level of granularity is sufficient to describe quality-related relationships [217].

The second important stakeholder is the *manufacturer* of the vehicle, the *OEM*. The manufacturer aims at low costs for designing and building but also using the system. The concern is mainly in two directions: (1) reuse of proven hardware from the last series and (2) power consumption. The former is an OEM concern, because it allows decreased costs and ensures a certain level of reliability which in turn also reduces costs by less defect fixes. The power consumption is an important topic in automotive development because of the high amount of electronic equipment that needs to be served. Hence, to avoid a larger or even a second battery – and thereby higher costs – the power consumption has to be minimised. Therefore, the relevant activities of the OEM are (1) *system integration* in which the software is integrated with the hardware and (2) *defect correction* which includes callbacks as well as repairs because of warranty.

For all parts of the instrument cluster, comfortable *driving* is still a demand. The *driver* should not be distracted in the *driving* activity. Hence, it must be safe. The most information can not surprisingly be found about the *TICS dialog* itself. It is often stated that it should be possible to obtain information and that the dialog should be "attractive" for the *driver*. The information displayed needs to be correct, current, accurate and authentic. In general, it is also stated that the dialog needs to be "well known" what we called "traditional", but it must also improve over the current systems. The *defect correction* should be minimal with a high robustness and lifespan. Finally, *system integration* should have minimal hardware requirements and use existing hardware components. The integrators should also be able to use different hardware, especially in the case of different radio vendors.

For further requirements analysis, we rank the identified activities of the two relevant stakeholders according to their importance for those stakeholders. The decisive view is usually the one from the project lead from the customer, in this case the *OEM*. Only legal constraints can have a higher priority.

Although we do not know how an automotive manufacturer would prioritise these activities, we assume that usually the safety of the driver should have the highest priority. Hence, we rank the *driving* activity above *defect correction* and *system integration*. The rationale for ranking *defect correction* higher than *system integration* is that the former is extremely expensive, especially in case the system is in the field already. This is partly backed up by the commonly known fact that it is more expensive to fix a defect, the later it is detected [200].

We can again use the quality model from [217] for most of the entities tree. It provides a decomposition of the *vehicle* into the *driver* and *TICS*. The *TICS* is further divided into *hardware* and *software*. The software is decomposed based on an abstract architecture of user interfaces from [217]. The hardware is divided into operating devices, indicators/display and the actual TICS unit.

The quality model gives us also the connection from activities to those entities. It shows which entities need to be considered for the activities of the stakeholders that are important for our instrument cluster. We cannot describe this completely for reasons of brevity but give some examples for refinements using the entities tree.

For the *driving* activity, we have a documented influence from the the hardware, for example. More specifically, the appropriateness of the position of the display has an influence on *driving*. We can express that in a compact way using the notation, we introduced in Sect. 2.4:

$$[\mathsf{Display.Position}\mid\textsf{APPROPRIATENESS}] \xrightarrow{+} [\mathsf{Driving}]$$

Hence, to reach the quality goals for the *driving* activity, we need to ensure that the display position is appropriate.

A second example starts from the *processing* activity. It is influenced by the unambiguousness of the representation of the output data:

$$[\mathsf{Output\ Data.Representation}\mid\textsf{UNAMBIGOUSNESS}] \xrightarrow{+} [\mathsf{TICS\ dialog.Processing}]$$

Therefore, we have a requirement on the representation of the output data that it must be unambiguous, i.e. the driver understands the priority of the information.

Finally, *perception* is an activity in the *TICS dialog* that is influenced by the adaptability of the output data representation:

$$[\mathsf{Output\ Data.Representation}\mid\textsf{ADAPTABILITY}] \xrightarrow{+} [\mathsf{TICS\ dialog.Perception}]$$

The representation should be adapted to different driving situations so that the time for *perception* is minimised. Such a requirement is currently missing in the specification [30].

For the specification of the requirements, the quality model can only help if there are measures defined for the product factors. We then can define an appropriate value for that measure. Otherwise, the model must be extended here with a measure, if possible. We can again not describe all necessary measurements of the instrument cluster specification but provide some examples.

The identified requirement about the appropriateness of the display position can be given a measure. The specification [30] demands that "The display tolerance [...] amounts to $\pm 1.5$ degrees". Furthermore, it is stated that "The angle of deflection of the pointer of the rev meter display amounts to 162 degrees".

We cannot describe the unambiguous representation of the output data with a numerical value. The specification [30], however, demands that the engine control light must not be placed in the digital display with lots of other information "because an own place in the instrument cluster increases its importance".

We could now extend this to specify further quality requirements. We summarised the example requirements in Table 3.5.

### 3.1.4 Checklist

- Have you thought about all stakeholders (end user, customer, developer, product manager, administrator, supplier, institutions)?
- Have you defined at least one goal per stakeholder?
- Have you understood the needs and goals of the stakeholders?
- Have you decided on for which goals you want to build the quality model?
- Have you collected all material that might be relevant for the stakeholder goals (process documentation, usage scenarios, standards, user documentation)?
- Have you identified the relevant activities for all stakeholders?
- Have you formed the activities into a hierarchy?
- Have you defined for each activity how well it needs to be supported to achieve the general goals?
- Have you identified all technologies that play a role in your product (programming language, middleware, database system, user interface, operating system)?
- Have you identified all relevant artefacts (requirements specifications, architecture specifications, models, source code, test cases)?
- Have you analysed all relevant material for the technologies and artefacts (specifications, standards, empirical studies, internal and external checklists and guidelines)?
- Have you defined at least one property for each relevant artefact to form a factor?
- Have you refined the stakeholder goals to concrete properties of the software product?
- Have you defined at least one impact from a factor to a relevant activity?
- Have all relevant activities impacts from at least one factor?
- Have you specified a quality requirement for each factor?

### 3.1.5 Further Readings

- Tian [198]
  There are not many sources that discuss quality planning for software. One good exception is this book. It discusses how you can map quality goals to metrics in software quality assurance. It provides you with a description from a slightly different angle but without including quality models.

- Chung et al. [38]
  This book describes the NFR (non-functional requirements) framework, which gives you a comprehensive method and guideline for handling this kind of requirements.
- Ficalora and Cohen [64]
  This book does not aim at software development, but it introduces well the concepts of QFD and Six Sigma as well as their combination. It is worth reading if you want to add ideas from these approaches to your quality planning and control.
- Lochmann [137]
  This paper gives an overview of how to engineer quality requirements using quality models based on the Quamoco quality meta-model.
- Plösch, Mayr and Körner [177]
  This paper describes a similar goal-based approach based on Quamoco-style quality models but adds the notion of *obstacles* to help in the refinement from goals to requirements.

## 3.2 V&V Planning

Verification denotes checking if a result fits to its requirements specification and validation describes checking a result against the stakeholder expectations. In brief:

Verification Have we done the product right?
Validation Have we made the right product?

Both activities in combination (abbreviated by V&V) make sure that we know the quality of our software product. There are many techniques that we can use to perform these tasks, such as reviews, tests or static analysis. You can find a detailed discussion of various V&V techniques in Chap. 4. As the effort for them can constitute up to 50 % of the initial development effort, it is necessary to plan and optimise V&V.

Collofello [42] expounds the general approach to V&V planning: Different organisational entities share the responsibilities for V&V depending on the company's organisation. The development team itself performs initial V&V techniques on their work products. For example, they review the specifications or write unit tests for their code. In addition, there should be an independent test team which tests and analyses further without being biased by the development team. A separate quality assurance team can be useful for additional tasks such as evaluating internal consistency or tracking quality metrics. Finally, an independent V&V contractor can add even more objectivity as well as potentially other test techniques and methods.

I will present a step-by-step approach to planning these V&V techniques independently of the actually used techniques.

### 3.2.1 *Step by Step*

Identification of V&V Goals

We discussed how to identify the quality goals and requirements in Sect. 3.1. They are the basis for the V&V goals. We need to identify which of the quality goals are feasible to be checked by V&V and then specify them as V&V goals. Moreover, we add constraints of our project such as schedule and effort restrictions that we need to consider in the further planning.

Selection of V&V Techniques

The selection of V&V is an optimisation problem with many factors and a lot of uncertain data. From the goal identification, we have the V&V goals for which we need suitable techniques, i.e. defect detection techniques that are able to check these goals. We also need techniques that are suitable for all the artefacts that we have to analyse, such as requirements specifications, design documents, code or test cases. Finally, the available techniques need different levels of effort and tool support. Although there are some academic proposal to systematise this selection, in practice, it is still mostly based on expert decisions, possible supported by data on defect detection and effort from old projects. As we base all quality control activities on a quality model, if the model is detailed enough, such as Quamoco quality models, we can derive the necessary techniques at least in part. It describes what measurements we need to make a quality evaluation and also how we perform these measurements. Therefore, it gives us the needed analysis tools and aspects to be reviewed or tested.

Integrating V&V Techniques

After we have identified a suitable set of techniques, we need to integrate them by distributing them over the life cycle phases of the product, deciding on which techniques are done in which order, and by assigning them to the responsible teams. This is especially demanding in a continuous quality control setting. We will do several V&V techniques not only once but several times, some even daily or hourly. Hence, we need to carefully plan what V&V techniques can be done in what frequency. For example, static analysis or certain automated tests lend themselves to a daily execution. Manual inspection or tests will probably only be performed at important milestones.

Problem Tracking

It is not sufficient to just perform techniques to check artefacts and detect defects, but the detected problems need to be tracked until they are resolved. If you do not closely track problems, they will be forgotten and appear again at the customer. You need to add information to the problem such as when and where it occurred, what preconditions are necessary and who is responsible for resolving it. There are many tools available that help in making this quick and clean. The open source bug tracker *Bugzilla*,[4] for example, supports developers and testers by providing a useful interface through their browsers. On top of that, you can track all changes to your product, not only problems. You can handle feature requests from customers in the same tool and then easily compare how much time and effort goes into what.

#### Tracking V&V Activities

Tracking the time and effort spent for performing V&V has several benefits. First, we can use the data from here in future projects as basis for selecting suitable techniques. Effort and defect detection capabilities vary strongly over different domains and teams. Hence, for an accurate prediction it is best to use data from the same team. Second, in some domains it is necessary to document V&V as part of the qualification of the product. For example, safety-critical software in the avionics domain needs to document their code inspections and test executions. Third, it helps the project manager to get an overview of the current state of V&V and thereby adjust schedule and budget. For this, we should not only track the effort and defects but also how much V&V is still necessary, e.g. how many tests still need to be performed.

#### Assessment

Finally, you should also include in the plan to evaluate the quality of the product as well as the process. The product quality evaluation is part of the quality control loop (Sect. 4.2). The assessment of the V&V process is needed to continuously improve it by removing unnecessary steps and refining the decision processes (Sect. 4.6). These evaluations then go into the assessment of the V&V planning: Have we chosen the right V&V techniques for our context? If not, it gives us hints for improving our planning for the next time.

### 3.2.2 Example: Instrument Cluster

To illustrate V&V planning, we come back to our example from Sect. 3.1.3: the instrument cluster. We hypothetically extend it to come to a complete V&V plan.

[4]http://www.bugzilla.org/.

The *Orthogonal Defect Classification* (ODC) [35, 36] is a popular approach for root cause analysis in software engineering. It classifies the defects you find in your quality assurance activities or later in the field. You can use it to get an overview of the types of defects you find in your products and what V&V techniques are best suited to find specific types of defects. Hence, it helps you in planning V&V by fitting techniques to probable defects.
A classification of a defect consists of two parts: the cause and the effect. In the effect, we describe the impact the defect has in terms of severity or reliability. For the cause, we choose a defect type and defect trigger. The trigger is the V&V technique that found the defect. You can determine the defect type by looking at what the developer does to fix the defect. A defect type is either an *incorrect* or *missing*

- Function
- Interface
- Checking
- Assignment
- Timing/serialisation
- Build/package/merge
- Documentation
- Algorithm

Although there is more information you can track for ODC, these are the most important ones. They give you all the information you need to plan and prioritise V&V techniques.

**Sidebar 3:** Orthogonal Defect Classification

### Identification of V&V Goals

The specified requirements constitute a part of our V&V goals. In addition, we have effort restrictions, such as a maximum effort of two person/month and an overall V&V schedule of three month. The defined goals are mostly possible to check by reviews of the design and the final system. Some of the goals might be easier to check on the activities level for which we employ user tests.

### Selection of V&V Techniques

We chose a usability review of the design and the final system as most comprehensive and cheaper as extensive user testing. To support other stakeholder goals and activities, e.g. the system integration, we select unit and integration tests as well.

### Integrating V&V Approaches

The design review does not depend on finished code or finished system but only needs a finalised design specification. Therefore, this is the first used V&V technique after the design phase. While the system is implemented, we continue with module tests so that the basic functionality is checked and the further V&V is not slowed down by simple defects. These will be done with the nightly build during the implementation phase. We can include integration tests in parallel to the finalisation of the modules. Hence, the integration test can be part of the continuous integration of the system. After the developers fixed module and integration defects, we plan the usability review of the final system.

#### Problem Tracking

For tracking all defects that are detected by V&V we set up an instance of *Bugzilla*.[5] All developers, reviewers and testers get an account and can post bugs as well as work on them.

#### Tracking V&V Activities

We use Bugzilla also for tracking the effort for the removal of bugs, the found defect types as well as which activity detected it. We use the existing time accounting system to document the effort for the detection activities.

#### Assessment

After the release of the system, we look at the tracked effort and defects and compare that to our initial planning. We will use that information to optimise V&V for the next iteration of the instrument cluster.

### 3.2.3 Checklist

- Have you searched for all possible V&V techniques for your product and its quality profile?
- Have you made a link from each stakeholder goal to a V&V technique that checks it?
- Have you systematically selected and ordered the V&V techniques?
- Do you have an explicit and tool-supported problem tracking in place?
- Do you track the spent effort and results of your V&V activities?
- After V&V: Have you compared your plan to the actual V&V and evaluation results?

[5]http://www.bugzilla.org/.

### *3.2.4 Further Readings*

- Wallace and Jujii [215]
  This is an old but still valid paper about what you can do in software V&V.
- Chillarege [35]
  This chapter of a handbook is in our opinion the best introduction to ODC. You can find it freely available on the Web.

# Chapter 4
# Quality Control

To counter quality decay during software evolution, proactive countermeasures need to be applied. This part of the book introduces the concept of continuous quality control and explains how it can be applied in practice. Particularly, it discusses the most relevant quality assurance techniques.

## 4.1 Quality Control Loop

The idea of continuous improvement processes as the basis for quality management systems originated in non-software disciplines in the 1980s [52,114,188]. It is now a standard approach for process-oriented quality standards such as the ISO 9001 [91]. A popular model for continuous improvement is the *PDCA cycle*, also called *Deming cycle* or *Shewhart cycle*. The abbreviation PDCA comes from the four phases of the model: Plan, Do, Check and Act. They are positioned in a loop as in Fig. 4.1.

In the *Plan* phase, we identify improvement potential, analyse the current state of the process and develop and plan the improvement. In the *Do* phase, we implement the process improvement in selected areas to test it. In the *Check* phase, we evaluate the results of the *Do* phase to be able introduce the improvement process in the *Act* phase. As you realise, the PDCA cycle is very general and can be used for software processes as well (see Sect. 4.6).

### *4.1.1 Product Quality Control*

Not only processes are in need of continuous improvement, however. Various researchers have shown that software ages and undergoes a quality decay if no countermeasures are taken [16, 170]. At first sight, this insight is counter-intuitive,

S. Wagner, *Software Product Quality Control*, DOI 10.1007/978-3-642-38571-1_4,

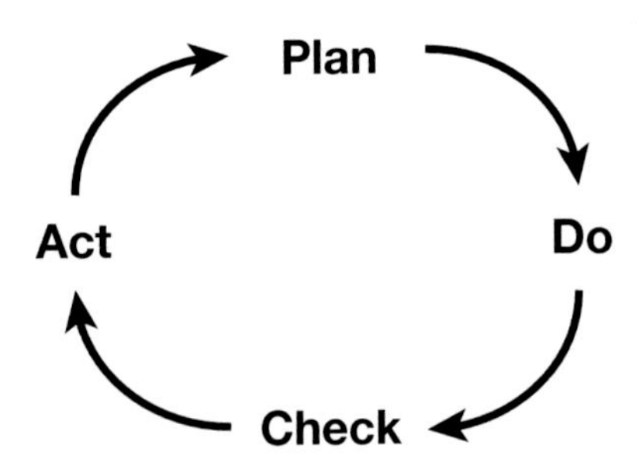


**Fig. 4.1** The PDCA cycle

because software is intangible and, hence, not subject to physical wear and tear. A 10-year-old software does not work worse because its bits eroded over time. Nevertheless, an ageing process takes place also for software.

All successful software system change during their lifetime. This is commonly called *software evolution*, although this term is counter-intuitive as well. Software changes are not random mutations from which the best survive. Quite the contrary takes place: The software engineers perform maintenance to fix existing defects, to restructure the system, to adapt or port it to new or changed platforms and to introduce new features. This is not random but very deterministic. These many changes to the system are necessary so that the system stays valuable for its users, but they are also a danger to the quality of the system.

In practice, owing to time and cost pressure but also missing training, changes are done too often carelessly of the effects on the quality of the system. For example, developers implement calls to other components that were not allowed in the original software architecture to get the system running on a new operating system version. New features that are similar to existing ones are implemented by copying the source code for the existing systems and making changes to the copies instead of introducing common abstractions. Algorithms are introduced because they were the first to come to mind not because they are the most efficient. Wild hacks are introduced to make changes work because nobody understands the original code anymore.

Hence, this affects all quality factors: buggy changes reduce reliability and bad algorithms reduce performance efficiency. The most stroke quality factor, however, is maintainability. The source code and other artefacts become more and more incomprehensible and a nightmare to change. At the same time, this is far less visible to management, because it has no immediate effect on the customer-perceived quality. To counter this problem, continuous improvement and control is required not only on the process but also on the product level. This enables us to proactively work against the otherwise inevitable quality decay.

Therefore, we proposed to use the same analogy of the PDCA cycle to describe the quality control process for software quality: *continuous software quality control* [48]. Figure 4.2 systematically shows this control process as a feedback loop. Starting from general product goals, we use a quality model, our central knowledge base about product quality, to specify quality requirements. Then, we give them

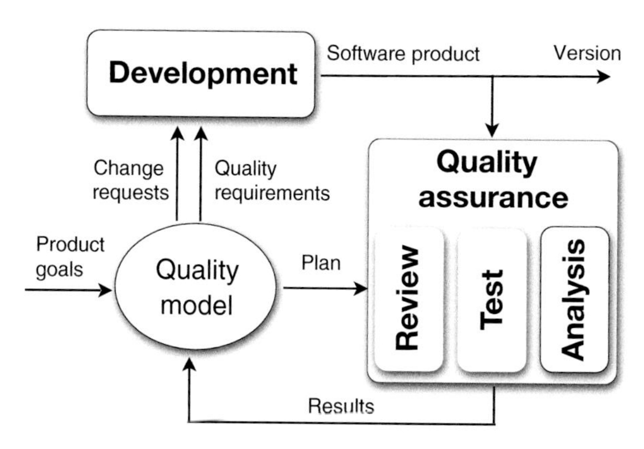


**Fig. 4.2** Quality control loop

to the development. They build a version of the software product that is given to the quality assurance team. They apply various quality assurance techniques based on what the quality model proposes to measure the specified quality requirements. This includes reviews, tests and other analyses. The results are fed back into the quality model and compared with the quality requirements. Deviations together with changed product goals lead to change requests for the development team who produce a new version of the product. The cycle continues.

Ideally, we perform this cycle as often as possible, for example, together with the nightly build and continuous integration. As not all of the quality assurance techniques are automatic, however, the cycle time of the complete feedback loop cannot be 1 day. We need to find a compromise between the effort to be spent for performing quality assurance and the gain from avoiding quality decay. Therefore, there need to be different cycle times depending on the quality assurance techniques applied in that cycle. Many static analyses can be performed autonomously and comparably fast so that they can be executed at least with the nightly build or even at every check-in. The execution of automated tests should be part of the continuous integration but long-running tests should also only be executed with lower frequency. Manual analyses, such as code reviews, are only feasible at certain milestones. The milestones should contain defined quality gates that describe the expected level of quality so that the project achieves the milestone. For such important points in the project, the corresponding effort should be spent.

Practice shows that two different cycle frequencies with different analysis depths work best to prevent quality decay without decreasing development productivity (Fig. 4.3):

1. We execute automatic quality assurance on a daily basis, e.g. during the *nightly build*. We summarise the results of these analyses on a quality dashboard (see below). This dashboard provides all project participants (including the quality engineer and project managers) with timely information and gives developers

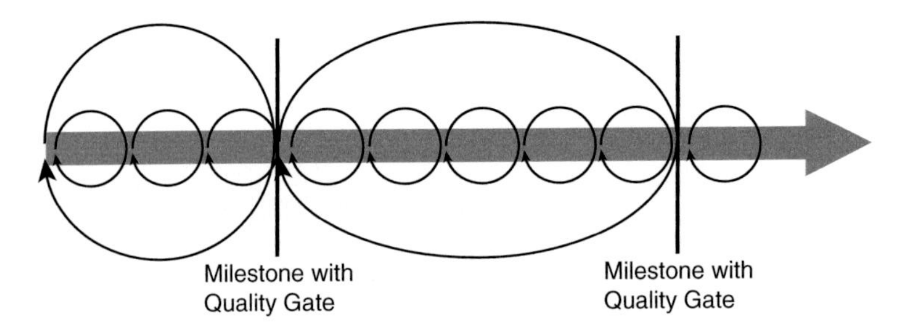


**Fig. 4.3** Two cycle frequencies

the opportunity to identify and correct quality defects immediately. Trend data provided by the dashboard supports the quality engineer in identifying negative trends over time.

2. In a second cycle, we execute higher effort quality assurance techniques roughly every month, at least at each important milestone with a quality gate. These quality evaluation uses the data provided by the dashboards but includes additional analyses that require manual work. We investigate changes in trends to ensure that reasons for improvement or decline are fully understood. These detailed quality evaluations should result in quality reports that summarise the findings of the evaluation and illustrates general trends. These reports serve a reliable documentation of the status quo that can be communicated to management and used as reference for future discussions about trends in quality.

### *4.1.2 Tool Support and Dashboards*

We have already discussed that automation is key for making continuous quality control a success. Only if we have a high level of automation, the higher quality control frequencies are possible. Hence, we need tool support. This includes:

1. Tool support for executing the actual quality assurance techniques, such as static analysis tools or automated tests.
2. A tool to combine the results and present to us the results, ideally already interpreted as a quality evaluation.

The probably most intensively used tool support for quality assurance today is automated tests. Not only unit test frameworks have made automating them highly popular but also other automation possibilities, such as capture/replay tools to automate GUI testing, are becoming popular. All these test automation tools are helpful for supporting quality control. As a result, we get the numbers of successful and failed test cases as well as test coverage. We will discuss testing in detail in Sect. 4.5. Static analysis is the second area that lends itself to automation. There is some tool support for manual reviews which can help to collect data about it. Even with this tool support, reviews will probably not be conducted with high frequency. Automated static analysis, however, works perfectly with high frequency in quality control. Bug pattern tools or architecture conformance analysis tools can deliver warnings about various problems; metrics tools and clone detection tools can collect measurements that give us a better understanding of the state of the quality of our software. We will discuss static analysis in detail in Sect. 4.4.

For combining all these singular results, we usually apply tools called *dashboards*. Similar as in other areas, these dashboards have the aim to provide us with an overview of our product or project. A dashboard does not have to be about quality alone, but we will concentrate on quality dashboards here. The three main tasks of a dashboard in this context are *integration*, *aggregation* and *visualisation*.

With integration, we mean that the dashboard is a central tool that is able to start the execution of other tools so we do not have to do that manually. Ideally, the dashboard is a kind of command-line tool so that it is easy to integrate in our regular build and continuous integration systems. We have used the dashboard tool ConQAT[1] in most of our projects, because people involved in these projects develop this tool. It is freely available and very powerful. It allows a high level of customisation and, thereby, it integrates test tools such as JUnit or static analysis tools such as FindBugs. Some analyses, such as clone detection, are directly implemented in ConQAT.

The second task, the dashboard should support, is the aggregation of the results of the various sources of data, i.e. tests, static analysis and manual reviews. Depending on the data we receive, it is either already system-wide or specific for modules or classes. Module-specific measures need to be aggregated to higher levels, such as packages and system. The dashboard should support to aggregate to different levels and allow the users of the dashboard to choose the level they are most interested in. For example, when we measure the fan-in and fan-out of a class using ConQAT, we then want to see average values for the package level and system level. We introduced the possibilities of aggregation in Sect. 2.2.2. In addition, it is helpful if the dashboard can already automate the interpretation of the measures in some kind of quality evaluation. In the Quamoco approach to quality evaluation (Sect. 4.2), we generate a suitable configuration for ConQAT to do that.

Finally, the dashboard should not only create numbers from the data sources and aggregation but support the dashboard users by visualising the outcomes. Even if we employ a very structured, goal-driven measurement program, the results can be huge. For large systems, it is easy to get lost in the amount of data we get. Humans are visual beings, and visualisation helps to get an overview and to find problems.

A visualisation that often can be a crude representation of the data but which is useful for a very quick overview is traffic lights. Formally speaking, it is usually a rescaling aggregation. Some complicated data in interval or ratio is aggregated into an ordinal scale of red, yellow and green. Green usually represents the values that are "good", i.e. on an accepted level, yellow are problematic values and red are

[1] http://www.conqat.org

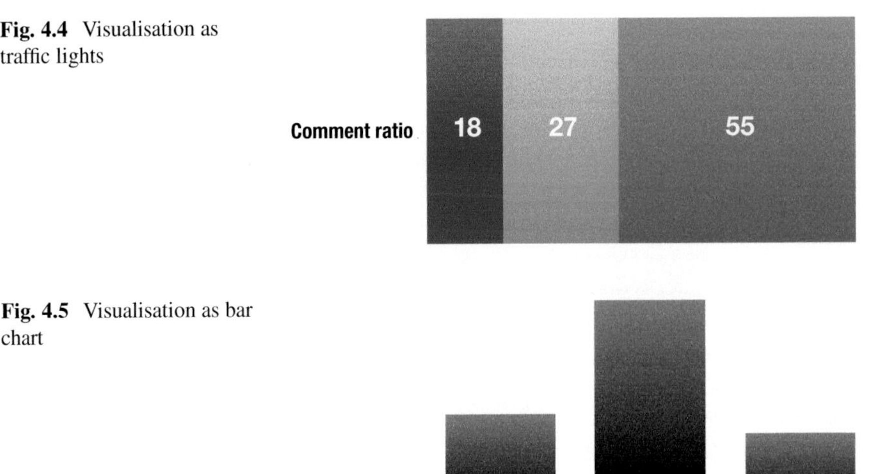


**Fig. 4.4** Visualisation as traffic lights

**Fig. 4.5** Visualisation as bar chart

"bad" values. The example in Fig. 4.4 shows the percentage of classes in a software system with comment ratios on the red, yellow and green level. Red can be defined as a comment ratio between 0 and 0.1, yellow as 0.1–0.2 and red above 0.2. Another use of traffic lights with only red and green is to show the percentage of unit tests that passed and failed.

A bar chart is a useful visualisation for comparing different values. The example in Fig. 4.5 compares the number of change requests for each type of change request. This shows us whether we mostly fix bugs (corrective) or if we are able to build new features (perfective) and restructure and improve the internals of the system (preventive).

Trend charts are visualisations that help to track measures over time. In many cases, I am interested not only in the current state of my system but also in the change over time. This is especially useful for measures that indicate problems, such as clone coverage or warnings from bug pattern tools. Often it is not possible and also not advisable to make immediate huge changes to the system to remove all these problems. Nevertheless, it makes sense to observe these measures over time and to make sure that at least they do not increase or even better that they decrease. The example shown in Fig. 4.6 is the percentage of clone coverage of a system over 6 month.

Finally, tree maps are mostly used for finding hot spots in the system. The intensity of the colour indicates the intensity of the measure. The example in Fig. 4.7 is a tree map for clone coverage created by ConQAT. As this is a probability, it is in the range 0–1. The visualisation is then white for 0 and completely red for 1. Any shades in between show different levels of clone coverage. The tree map has the additional dimension of the squares it shows. These squares can show another measure, usually the size of classes or components.

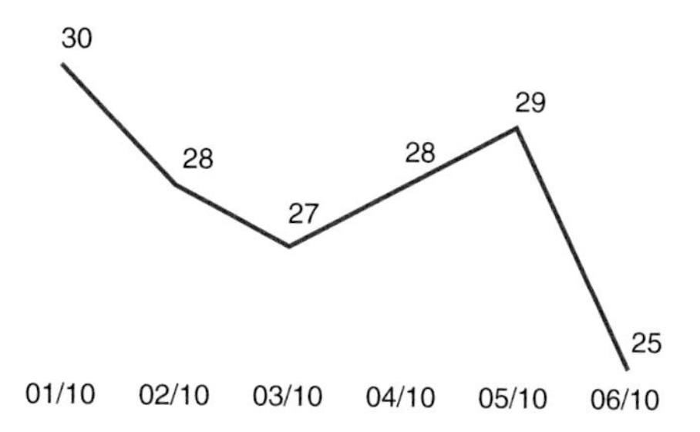


**Fig. 4.6** Visualisation as trend chart

**Fig. 4.7** Visualisation as tree map

### *4.1.3 Summary*

In summary, we apply continuous quality control for the product quality of our software systems to counteract on the otherwise inevitable quality decay. We perform various quality assurance techniques in their respectively appropriate cycle frequencies to get a continuous feedback on the quality level of our product and are therefore able to react in a timely manner. Automation is key to make this quality control continuous, and especially a dashboard tool can help us to integrate, aggregate and visualise the results. How we perform the necessary analyses and quality evaluations is the subject of the remainder of this chapter.

## 4.2 Quality Evaluation and Measurement

The objective of the Quamoco quality modelling approach (see Sect. 2.4) was to make quality more concrete. We wanted to break it down to measurable properties to get a clear understanding of the diffuse quality characteristics we usually have to deal with. The resulting quality model structure allows us to do this now: collect data to measure product factors and aggregate the results up to quality aspects. Why do we need such a quality evaluation? It gives us a variety of possibilities: We can verify if our product conforms to its quality goals. We can find problematic quality aspects and drill down in the results to find improvement opportunities. We can compare different systems or sub-systems, for example, when we have to select a product to buy. We could even use this as the basis for certifying a certain level of quality. Figure 4.8 shows an overview of how this quality evaluation is performed.

### 4.2.1 Four Steps

Conceptually, there are only four steps for getting from the software product to a complete quality evaluation result:

1. Collect data for all the measures from the software product.
2. Evaluate the degree of presence of the product factors based on the measure values on a scale from 0 to 1 for each product factor.
3. Aggregate the product factor evaluations along the impacts to quality aspects using weights.
4. Aggregate further along the quality aspect hierarchy and interpret the results using a mapping to grades.

What do we need in the quality models to make this work? We have already discussed measures and their concrete measurement descriptions for specific programming languages or technologies in instruments. Therefore, we can use them directly to collect values for the measures. In addition, there is a model element called *evaluation* for each factor, i.e. all product factors and quality aspects. In the evaluation we can describe how we transform the results from the lower levels into an evaluation of the factor. Hence, for product factors, the evaluation describes how the values for its associated measures are combined to come to its evaluation. For quality aspects, the evaluations of lower-level product factors and/or quality aspects are aggregated to an evaluation of that quality aspect. In principle, these evaluation specifications could only be textual, giving a description for a human to make the evaluation. As there are hundreds of such evaluations in realistic quality models, we need to automate that. Therefore, we have machine-readable evaluation specifications.

For product factors, we call them *measure evaluations*, because they specify the calculation for getting the product factor evaluation from the values of the measures. Let us look at the example of the product factor [Source Code Part | DUPLICATION].

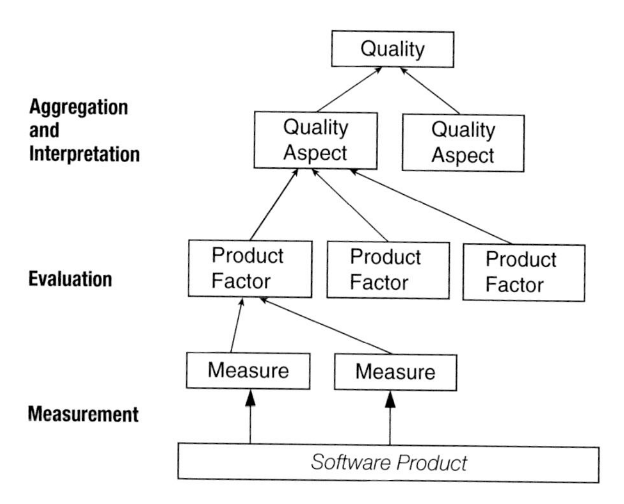


**Fig. 4.8** Overview of the evaluation approach [211]

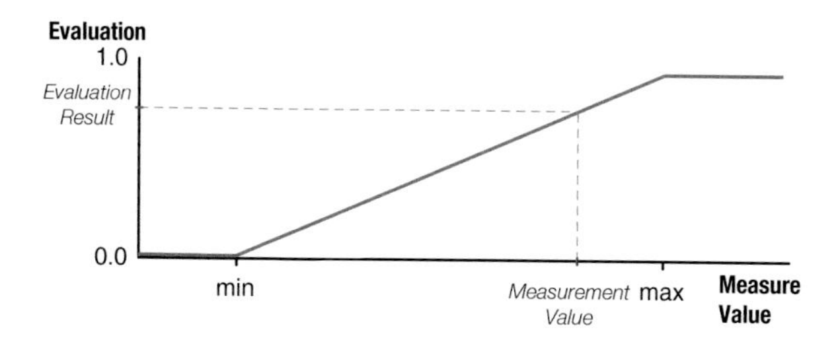


**Fig. 4.9** Example of a linear evaluation function [211]

It describes that "source code is duplicated if it is syntactically or semantically identical with the source code already existing elsewhere". We concentrate on measuring the syntactically identical code with the measure *clone coverage*. It describes the probability that a randomly chosen code statement is duplicated in the system. Hence, the higher the clone coverage, the more the product factor [Source Code Part | DUPLICATION] is present in the system. One could now start to assign different levels of clone coverage to the evaluation or just directly calculate it by the equation "1−clone coverage". To make such calculation more uniform and to express that there are usually two thresholds – a lower acceptable boundary and an upper acceptable boundary – we use special linear functions. Figure 4.9 shows an example.

The intuition is that there are two thresholds. Above and below these, I do not care about the detailed measure value, but the product factor is completely present or not present at all. Between these threshold values (min and max), the evaluation increases linearly. For the [Source Code Part | DUPLICATION] example and the measure *code coverage*, this means that below a certain threshold, say 0.05, I judge that there is no duplication in the source code. Above the max threshold, say 0.5, I consider the source code completely duplicated. Between min and max, the evaluation increases. Therefore, I can find for a certain measurement value for a concrete system an evaluation result between 0 and 1. Depending on the measures and product factors, the linear function could be the other way round.

This is rather easy for *code coverage*, because it is already a measure on the system level, i.e. it describes a property of the whole system. We have many other measures, however, that are not system-level measures. We use many rules from static analysis tools which single findings. We can sum up these findings, but it is still not the same if I have 10 findings in a system with 1,000 LOC or 10,000,000 LOC. What we need is a *normalisation* of the measurements with respect to the system. For that, we either set the LOC of the affected system part in relation to the LOC of the complete system or the number of affected entities in relation to the number of entities. A finding about missing constructors in classes can be summed and divided by the total number of classes in the system. A finding about a more specific library method, such as a missing closing of a database connection, is easier normalised by LOC. We bring all measure to the system level so that we can combine them more easily in the measure evaluations.

The evaluation specification is simpler for quality aspects. A quality aspect has either a set of product factors that impact them, other quality aspects that refine them or a mixture thereof. We chose again a simple unification of these cases by using a weighted sum. Hence, each impact gets a weight that describes how much of the complete evaluation of a quality aspect is driven by this product factor or quality aspect. We then sum all weights to calculate the weighting factor for each product factor or quality aspect. We multiply each evaluation result with its corresponding weighting factor and add up all of these to the evaluation for the quality aspect. The only thing we need to consider is whether impacts from product factors are positive or negative. A negative impact means that we use 1−the evaluation result from the product factor. For example, there are many impacts onto the quality aspect *analysability*. Let us assume there are only three:

- [Source Code Part | DUPLICATION] $\xrightarrow{-}$ [Analysability], weight 10, evaluation 0.2
- [Source Code Part | USELESSNESS] $\xrightarrow{-}$ [Analysability], weight 5, evaluation 0.3
- [Source Code Part | DOCUMENTEDNESS] $\xrightarrow{+}$ [Analysability], weight 7, evaluation 0.5

Hence, the weighting factor is 22, the sum of the weights. For the evaluation result for analysability, we calculate $10/22 \cdot 0.2 + 5/22 \cdot 0.3 + 7/22 \cdot 0.5 = 0.32$. A simple sum adjusted by the weights gives us again the degree of presence of the quality aspect. Analysability is present to a third. This is a good first indication but hard to interpret: Is that good or bad?

### 4.2.2 Interpretation Schema

Therefore, we also introduced an interpretation schema for quality aspects. The intuition is that we judge software quality similar to a dictation in school. You have to write a text read out to you, and afterwards the teacher counts the mistakes. The

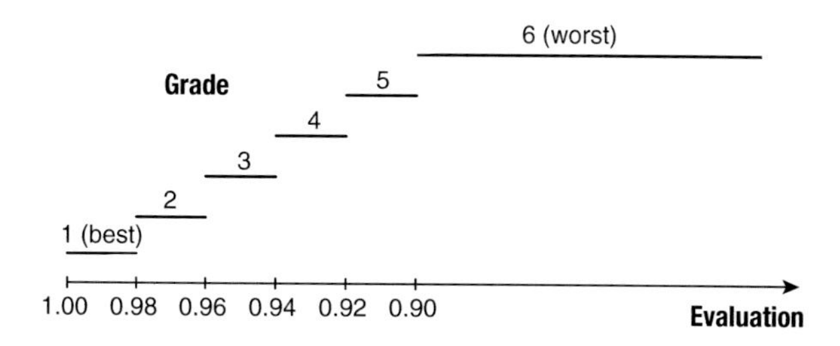


**Fig. 4.10** Interpretation as grades [211]

more mistakes, the worse the grade you get. If you want a really good grade, your text needs to be almost flawless. This is how we interpret the evaluation results between 0 and 1. An example interpretation using German school grades is shown in Fig. 4.10. In our previous example, the analysability with 0.32 would clearly result in a 6, the worst grade. You should definitely do something to improve it.

The final question, you probably have, is where we get the thresholds for the evaluation specifications and interpretations from. We call this *model calibration* because we calibrate these values usually according to empirical data. You can also calibrate each specification by hand using your own expert opinion. This is, however, difficult to repeat and to explain. Therefore, we rely on a more objective method: We apply the measures to a large amount of existing systems to understand what is "normal". The normal range defines then the upper and lower thresholds. You can find more details about the calibration in [211].

## 4.3 Walk-Throughs, Reviews and Inspections

Structured reading of artefact content is one of the most important quality assurance techniques as it enables one to detect problems that might lead to failures as well as problems that make it harder to comprehend the content. No other technique is as effective as reviews in finding this variety of problems. Depending on the exact execution of the reading, this technique is called walk-through, review or inspection. Unfortunately, it is not used frequently in practice. Ciolkowski et al. [39] found in a survey that less than half of the respondents perform requirements reviews and only 28 % do code reviews. Major obstacles stated were time pressure, cost and lack of training. To overcome these obstacles, the following gives you an overview of the available techniques, insights into the actual effectiveness and efficiency of reviews as well as suggestions on automation for reducing costs and integrating reviews into continuous quality control.

### *4.3.1 Different Techniques*

Because there are many variations of walk-throughs, reviews and inspections the terminology differs in existing literature and practical use. We give one possible

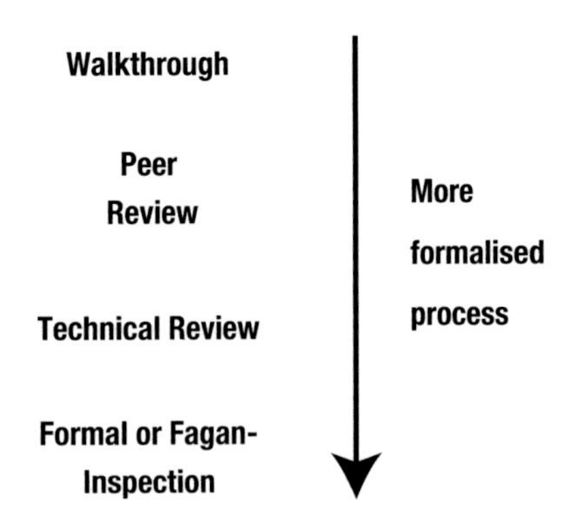


**Fig. 4.11** The main review techniques

terminology that fits to most usages of terms we have encountered, but you may find other uses easily. The most important differentiation between different review techniques is the formalisation of the review process. As Fig. 4.11 shows, walk-throughs are the least formalised, while formal inspections have the most formalised process. This means in practice that walk-throughs can be conducted in various ways. An inspection, in contrast, has clear process steps that have to be followed and that should be fixed in the company-wide development process.

The boundaries between these different techniques is blurry and one technique might borrow specific parts from another to make it more suitable for the task at hand. Most variation in these techniques can be explained along four dimensions that Laitenberger [132] identified: process, roles, artefacts and reading techniques. The process defines the steps of the review and which steps are mandatory. The participating roles determine who has to take which part in the review. The artefacts that are reviewed, including support artefacts such as checklists or requirements, prescribe when in the process it is possible to use the review, because the artefacts have to be created first. Finally, the used reading techniques differentiate how the reviewers conduct their reading which also has influences on the required artefacts.

*Walk-throughs*, also called presentation reviews, have the aim to give all participants an understanding of the contents of the analysed artefact. The author guides a group of people through a document and his or her thought processes to create a common understanding. It ends with a consensus on how to change the document. For that we need only two roles in the walk-through process: the author and listeners. The solely defined process step is the explanation of the document by the author. How the consensus is reached is usually left open. A walk-through is possible for all kinds of documents. Specific reading techniques are not used.

In *peer reviews*, the authors do not explain the artefact. They give it to one or more colleagues who read it and give feedback. The aim is to find defects and get comments on the style. Similar to walk-throughs, peer reviews make use of only two roles, the *author* and *peers*, usually colleagues that are organisationally close to the author. There are two process steps: (1) individual checking by the peers and (2) feedback from the peers to the author. The peers often give the feedback informally in a personal discussion, but they can also write it down. It is possible

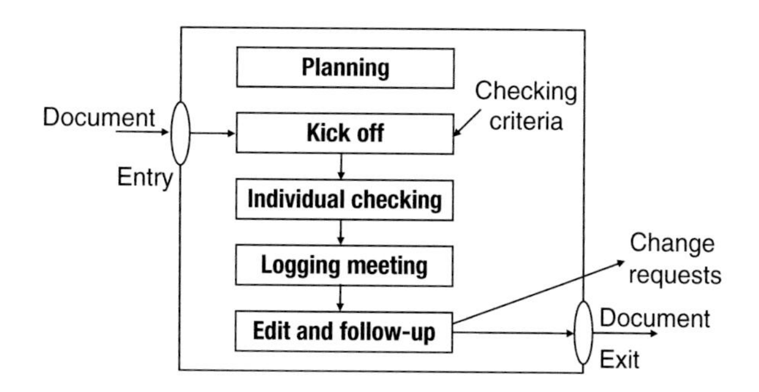


**Fig. 4.12** The inspection process (derived from [73])

to use different reading techniques in the individual checking step, but it is usually not demanded by the process. The author could ask the peers, however, to look for specific aspects. You can use a peer review for any kind of document.

*Technical reviews* further formalise the review process. They are often also management reviews or project status reviews with the aim to make decisions about the project progress. In general, a group discusses the artefacts and makes a decision about the content. The variation in the process and roles is high in practice. There are at least the two process steps individual checking and joint discussion. The involved roles depend on the type of artefact that is reviewed. For example, in a design specification, implementors as well as testers should be involved. Again, different reading techniques can be used, but most often, it is checklist-based reading.

*Inspections* have the most formalised process which was initially proposed by the software inspection pioneer Fagan [59]. They contain formal individual and group checking using different sources and standards with detailed and specific rules how to check and how to report findings. Furthermore, we distinguish between the roles *author*, *inspection leader*, *inspector* and *scribe*. The author, however, does not take part in the actual inspection but is only involved to answer questions and correct minor problems directly after the inspection. The leader is the moderator and organiser of the inspection. The inspectors read and report findings, while the scribe documents them.

The process of an inspection consists of the steps *planning*, *kick off*, *individual checking*, *logging meeting* and *edit and follow-up* (Fig. 4.12). In the planning step, the inspection leader, or moderator, organises a particular inspection when artefacts pass the entry criteria to the inspection. The leader needs to select the participants, assign them to roles and schedule the meetings. The kick off meeting, sometimes called overview step, is not compulsory but is especially useful if inspections are not well established or there are many new members in the inspection team. This phase is intended to explain the artefact and its relationships to the inspectors. It can be helpful for very complex artefacts or for early documents such as requirements descriptions.

The next step, individual checking, is the main part of the inspection. The inspectors read the document based on predefined checking criteria. Depending on these criteria, they need to use different reading techniques. The most common technique is checklist-based reading. The checklist contains common mistakes and important aspects that the inspectors need to look for. You can find, however, a variety of alternative reading techniques in the literature. Usage-based reading [196] proposes to employ use cases as a guide through the document, defect-based reading to define and then look for specific types of defects and perspective-based reading [9] emphasises reading from the point of view of different stakeholders of the document, e.g. developers, testers or customers.

The logging meeting is the next step in the inspection process. All inspectors come together to report their findings which are logged by the scribe. The author may take part to answer open questions of the inspectors. In some instances, here the document is read again together, but most commonly, only the found issues are collected. The main benefit of the meeting is that the collective discussion of findings reveals new issues as well as establishes a consensus on the found problems. The logging meeting closes with a joint decision whether the artefact under inspection is accepted or needs to be reworked. In the edit and follow-up step, the author removes minor problems directly from the document. For larger issues, the inspection leader creates change requests. This step ensures that the document conforms to the exit criteria. Inspections are possible for all kinds of documents.

### *4.3.2 Effectiveness and Efficiency*

In practice, reviews are often considered less important than testing [212]. This is in contrast, however, to the findings we have that reviews are effective and efficient in quality assurance. Although research has left some questions open, there is a considerable body of knowledge that you can make use of in your quality control loop. Unfortunately, the studies do not always clearly distinguish between walk-throughs, peer reviews, technical reviews and inspections. Most studies, however, investigate more formalised review techniques such as technical reviews and inspections.

We found by analysing a variety of studies and reports from practice that on average about a third of the defects in an artefact is found by reviews [200]. The variance of the effectiveness, however, is large. It can go up to 90 % if the review is conducted correctly. This means that there is a defined process and that the reading speed is about one page per hour. If we read less, we do not find more defects, but if we read more than two pages per hour, we miss defects.

Despite the large conceived effort needed for reviews – several people need to read the same documents – they are efficient. On average, you find 1–2 defects per person-hour including all preparation and meeting efforts. This is at the same level as testing. Similarly to the effectiveness, efficiency can be lifted even higher. High-quality inspections find up to six defects per person-hour which is hardly achievable with testing. Moreover, most studies show that tests and reviews find (partly) different defects [184]. Therefore, to find as many defects as possible before you ship the software to your customer, you need to perform reviews in addition to tests. This is also stated concisely in the Hetzel–Myers law [57]: "A combination of different V&V methods outperforms any single method alone".

Moreover, reviews are an extraordinary technique as they are generic and therefore very flexible in terms of when and on what you can use them. In particular, they have the huge advantage over most other quality assurance techniques that you can employ them early in the development on prose documents such as requirements specifications. Hence, you can find and remove the defects when it is still comparably cheap. Furthermore, in comparison to dynamic techniques such as testing, a review gives the exact defect localisation. A reviewer identifies a problem and can point the author to the exact place where it occurs. A tester can only show wrong behaviour. We found in the existing studies that the effort for removing defects found in a requirements review is on average one person-hour per defect, 2.3 person-hours per defect in a design review and 2.7 person-hours per defect in a code review. Hence, as rule of thumb, you have to spend 1 person-hour for a requirements defect, 2 person-hours for a design defect and 3 person-hours for a code defect. These results strongly encourage to perform a lot of reviews early in the development to save later on. As anything, this has a break-even point where it is not beneficial anymore to further invest into early reviews. To instantiate early reviews at all, however, is always beneficial.

Finally, the soft benefits of reviews are not to be underestimated. A review always means that different people from the development project read documents and then sit together to discuss them. Ideally, these people represent different roles and different levels of expertise. This way, the review also helps to create a common understanding between different stakeholders, the developers and the testers, for example. Furthermore, it serves to transfer knowledge from the more experienced team members to people new in the project or the profession. Nobody has been able to quantify this effect, but our experiences and those of our collaborates suggest that it is huge. If none of the data has convinced you to introduce reviews, this is a strong argument to do so.

### *4.3.3 Automation*

Automation is crucial for including such a strongly manual task as reviews in continuous quality control. Depending on the concrete review technique you use, different tool support is possible. Furthermore, different aspects of the review process can be automated. Tenhunen and Sajaniemi [195] identified four major areas that can be automated:

- Material handling (all document-related issues such as file format, versions or navigation)
- Comment handling (all comment-related functions, such as adding, viewing, reviewing, reporting or summarising)
- Support for process (features that automate or otherwise benefit the inspection process)
- Interfaces (use of external software or other tools)

Any review technique can benefit from automation in material and comment handling. Having both only on paper hampers quick analyses and distributed checking. In modern distributed development, it is often necessary to have reviewers around the globe that need quick access to the material. The tool-supported handling of comments allows the author to have quick access to the comments to incorporate them. If the tool supports further aspects of the process, it is necessary that these are clearly defined. Then, however, many benefits can arise such as automatic distribution of reviews to suitable reviewers or collection of review performance data that the project manager can analyse to improve the review process. These data then can also be fed into the continuous quality control process. This would be an interface to the corresponding tool support for quality control. But interfaces to further tools such as static analysis tools (Sect. 4.4) can also be beneficial. The warnings produced by the analysis tool can be shown to the reviewer in-line with the code to be reviewed. This helps the reviewer not to miss any problematic aspects.

For example, the Android Open Source Project[2] uses a web-based review tool for their peer reviews. They use Gerrit,[3] also an open source tool, that is based on the Git configuration and version management system. The philosophy in the Android project is that any change needs to be approved by a senior developer who checks out the change and tests whether Android still compiles and does not crash, i.e. the developer performs tests. The confirmation is handled within Gerrit. Moreover and more interesting for reviews, each change also has to be read by at least two reviewers. The list of approvers and reviewers for each change can be seen within the Gerrit web environment. The reviewers see the complete change set, i.e. all changed files, and can read the changes either in a side-by-side compare view or in one unified view. They have the possibility to add comments to any line in the changed files and start discussions about these comments. Finally, they can approve or disapprove of the changes. All team members get an immediate overview of all changes, comments and review tasks. Another example is the open source review manager *RevAger*,[4] which supports in contrast to Gerrit the traditional face-to-face logging meeting but automates many aspects around this meeting and is especially efficient in logging the found issues.

### *4.3.4 Usage*

Reviews are essential in the quality control loop to analyse factors in the artefacts that cannot be automatically assessed. This includes many aspects that have a semantic nature, i.e. that depend on the understanding what the content of the artefact means. For example, the appropriateness of code comments cannot be

---

[2]http://review.source.android.com

[3]http://code.google.com/p/gerrit/

[4]http://www.revager.org

assessed completely automatically. There are some automatic checks that give indications, but if you want to know whether the comment fits to what is done in the code, someone has to read both and compare them. Hence, there are many factors that have to be evaluated in the quality control loop that need reviews. For each work product in your development process, you should therefore plan at least a walk-through. Critical documents should be inspected.

Hence, reviews are indispensable to assess what you have specified in your quality model (see Sect. 2.4). Especially, if you invested in building detailed and concrete quality models following our Quamoco approach, you can and should exploit this by generating the checklists for your reviews directly from the quality model. The tool support we discussed in Sect. 2.4 is able to do that. This gives you two benefits: (1) the guidelines are streamlined with the quality goals and requirements in your quality model, and (2) you reduce redundancies between the quality model and the checklists.

Nevertheless, reviews need considerable manual effort. Therefore, they cannot be performed in a daily build. The frequency of reviews should be determined by other factors such as the finalisation of, or changes in, a specific document. For example, you should perform a review after you finished a requirements specification or the source code of a Java class. You should persist the information about quality problems that the review generated to use it in the quality control loop. Even if the information is not always collected freshly, it is available at any time. Furthermore, the information which parts of the system have been reviewed itself is interesting for quality control and should be collected and visualised.

### *4.3.5 Checklist*

- Have you made sure that reviews are done in an atmosphere of joint improvement instead of blaming the author?
- Is there a planned review (at least a walk-through) for any relevant document in your development process?
- Have you identified critical documents and scheduled a formal inspection?
- Have you made clear, for each review to be conducted, what type of review it should be, who has to participate and how it is going to be conducted?
- Do you make sure that the reviewers have access to all other documents that they might need during their checking (requirements specifications, checklists, design documents)?
- Have you made sure that the used other documents are derived from and are in sync with your overall functional and quality goals?
- Do you derive checklists automatically from your quality models to reduce redundancy?
- Do you adhere to the optimal reading speed of about one page per hour?
- Do you have a defined infrastructure that saves the results of the review as well as other data that was collected?

- Do you use tool support wherever possible in the review process?
- Have you incorporated the review results in overall (continuous) quality analyses?
- Have you modelled your existing checklists in a quality model and do you use this model for generating (role-specific) checklists?

### *4.3.6 Further Readings*

- Gilb and Graham [73]
  This is the classic book about inspections. If you are new to inspections and plan to introduce them to your development process, I highly suggest you to read it. It gives a lot of practical guidance and forms you can use. Because of its age, the book does not include up-to-date information about automation.
- Shull et al. [189]
  This is a gentle introduction to perspective-based reading, a specific reading approach, which you can use in your inspections. This article explains the approach and discusses experiences.

## 4.4 Static Analysis Tools

Static analysis tools are programs that aim to find defects in code by static analysis similarly to a compiler. The results of using such a tool are, however, not always real defects but can be seen as a warning that a piece of code is questionable in some way. Hence, the differentiation of true and false positives is essential when using static analysis tools. There are various approaches to identify critical code pieces statically. The most common one is to define typical bug patterns that are derived from experience and published common pitfalls for the programming language. Furthermore, the tools can check coding guidelines and standards to improve readability. Finally, more sophisticated analysis techniques based on the data flow and control flow find, for example, unused code or buffer overflows. We will give an overview of these techniques and discuss clone detection in more detail, because it proved very useful in our practical experiences.

### *4.4.1 Different Tools*

Any tool that does not execute the software for analysing it but uses the source code and other artefacts belongs to the class of *static analysis tools*. Hence, the differences between the tools at the ends of the spectrum are huge. The rough

**Fig. 4.13** Classification of static analysis tools

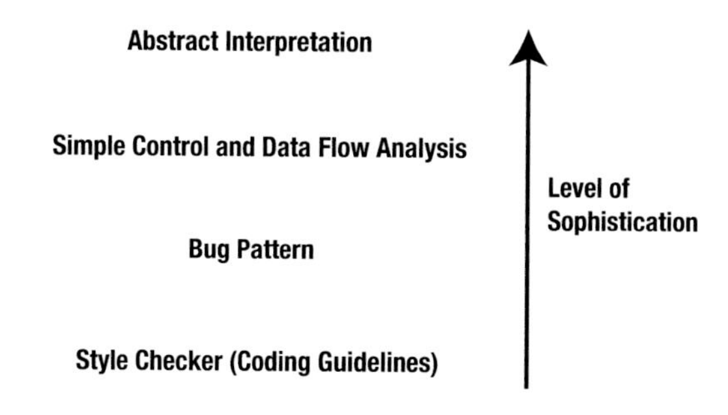


classification of Fig. 4.13 calls one end of this spectrum *style checkers*. These tools analyse the source code of a system for conformance with predefined coding guidelines. The analysis includes formatting rules, the casing of identifiers or organisation-specific naming conventions. These analyses are comparably simple and the tools can automatically resolve many problems they find. This functionality is basic enough so that integrated development environments (IDE) often contain this kind of checks. Eclipse, for example, can directly detect if the code does not conform to a defined format in the standard Java development edition. Another style checker for Java is Checkstyle[5] that in addition can check, for example, that you do not use more than a specified number of statements in a method or Boolean operators in an expression.

The most common type of static analysis tools is *bug pattern tools*. They formalise common pitfalls and anti-patterns for a specific programming language. For example, in Java, strings should be compared using the *equals()* method instead of ==. The latter would only check whether the compared variables point to the same object, not if they contain the same strings. Although in some cases the comparison of object IDs is the intended behaviour, in most cases, it constitutes a mistake by the programmer. This likely defect is found by a bug pattern tool which issues a warning to the programmer.

Most types of defects found by bug pattern tools are similar to the string comparison example. The bug patterns need to be described in purely syntactical terms; the semantics cannot be analysed. A run of such tools is quick and computationally cheap. Bug pattern tools are available for virtually any programming language. The availability, however, differs strongly: for Java, there are several usable open source tools such as FindBugs[6] and PMD[7], and for C, the prevalent tools in practice are mostly commercial such as PC-lint[8] and Klocwork Insight.[9]

[5]http://checkstyle.sourceforge.net/

[6]http://findbugs.sourceforge.net/

[7]http://pmd.sourceforge.net/

[8]http://www.gimpel.com/

[9]http://www.klocwork.com/products/insight/index.php

A more sophisticated analysis than merely searching for syntactical patterns is control and data-flow analyses. Bug pattern tools often contain a small number of detectors that use the control or data flow to find possible defects. For example, FindBugs can detect *dead code*, private methods that are never used, which is a control-flow analysis. PC-lint employs data-flow analysis to find variables that are written but never read.

Abstract interpretation generalises these simple control and data-flow analyses to any abstraction of the code that is abstract enough for an automatic analysis but concrete enough to make precise statements about properties of the program. Tools such as Coverity[10] or PolySpace[11] find buffer overruns or pointer accesses outside valid memory.

## 4.4.2 Clone Detection

One of the static analysis techniques I want to emphasise in particular is clone detection. It is a powerful technique to get an impression of the ageing process of a software, and it is highly automated at the same time.

### What Is a Clone?

A clone is a copy of a part of a software development artefact that appears more than once. Most of the clone detection today concentrates on code clones but cloning can happen in any artefact. In code, it is usually the result of a normal practice during programming: I realise that I have implemented something similar somewhere else. I copy that part of the code and adapt it so it fits my new requirements. So far it is not problematic, because we would expect now that the developer performs a refactoring to remove the introduced redundancy. Often, this does not happen, however, either because of time pressure or because the developer is not even aware that this can be a problem.

The developer does not keep an exact copy of the code piece but changes some identifiers or even adds or removes some lines of code. The notion of clones incorporates that. To identify something as a clone, we allow normalisation to some degree, such as different identifiers and reformatting. If complete statements have been changed, added or deleted, we speak of *gapped clones*. It is one of the parameters we have to calibrate in clone detection how large this gap should be allowed to be.

[10]http://www.coverity.com/

[11]http://www.mathworks.com/products/polyspace/

### Impact of Cloning

It is still questioned today in research if cloning is really a problem [118]. From our studies and experiences from practice, cloning should clearly be avoided because of two reasons.

First, it is undeniable that the size of the software becomes larger than it needs to be. Every copy of code adds to this increase in size which could be avoided often by simple refactorings. There are border cases where a refactoring would add so much additional complexity that the positive effect of avoiding the clone would be compensated. In the vast majority of cases, however, a refactoring would support the readability of the code. The size of a software is correlated to the effort I need to spend to read, change, review and test it. Any method cloned needs an additional unit test case which I either also clone or have to write anew. The review effort increases massively, and the reviewers become frustrated because they have to read much similar code.

Second, we found that cloning can also lead to unnecessary faults. We conducted an empirical study [113] with several industrial as well as an open source system in which we particularly investigated the gapped clones in those systems. We reviewed all found gapped clones and checked whether the differences were intentional and whether they constitute a fault. We found that almost every other unintentional inconsistency (gap) between clones was a fault. This way, we identified 107 faults in five systems that have been in operation for several years. Hence, cloning is also a serious threat to program correctness.

### Clone Detection Techniques

There are various techniques to detect clones in different artefacts [128]. Most of them are best suited to analyse source code but there are also approaches to investigate models [47] or requirements specifications [112].

We work mainly with the implementation of clone detection as provided in the tool ConQAT. It is proven as working under practical conditions, and it is now used at several companies to regularly check for cloning in their code. The measure we use to analyse cloning is predominantly *clone coverage* which describes the probability that a randomly chosen line of code exists more than once (as a clone) in the system. In our studies, we often found clone coverage values between 20 % and 30 % but also 70–80 % is not rare. The best code usually has single digit values in clone coverage.

In general, false positives are a big problem in static analysis. For clone detection, however, we have been able to get rid of this problem almost completely. It requires a small degree of calibration of the clone detection approach for a context but then the remaining false-positive rates are neglectable. ConQAT, for example, provides a blacklisting and can read regular expression to ignore code that should not be considered, such as copyright headers or generated code.

Finally, we use several of the visualisations the dashboard tool ConQAT provides to control cloning: a trend chart shows if cloning is increasing or a tree map shows us which parts of our systems are affect more or less strongly by cloning.

### 4.4.3 Effectiveness and Efficiency

In terms of ODC defect types (see Sidebar 3 on page 108), bug-finding tools are effective in finding two types of defects [220]: assignment and checking. Assignment defects are incorrect initialisations or assignments to variables. Checking defects are omissions of appropriate validation of data values before they are used. Despite all tools being effective in these defect types in general, different tools tend to find different specific defects because of their detection implementations.

Although the bug-finding tools are mostly reduced to these two defect types, they are able to find up to 80 % of the defects in a software product [200, 209]. The defects found are, however, mostly of low severity. For severe defects, the effectiveness is about 20 % [200, 209]. The most severe defects are of the types function and algorithm, i.e. constitute problems in the functionality and behaviour. These problems are very hard to detect using static analysis. In an industrial study, we could not find a single field failure that could have been prevented by using the static analysis tools [205]. In the same study, however, we found that code modules with a high number of warnings from different tools have a high probability to contain severe faults. Hence, even if the tools might not directly lead you to a fault that causes a field failure, you have a high chance to identify code modules that are badly written and hence contain such faults.

The effort needed to detect a defect is roughly in the same order of magnitude as for inspections [220]. A lot of the effort, however, goes into the installation and configuration of the tool. It is specifically important to adapt the tools to your environment, for example, your programming guidelines. With this adaptation, it is possible to reduce the false-positive rates to a tolerable level. Non-adapted static analysis can result in false-positive rates of up to 96 % [209]. After adaptation, the defect detection is quick and cheap and results on average in less than half a person-hour per defect [205]. On top of the effort, there can be additional licence fees for the tools which are significant for most commercial tools. Depending on the programming language, however, there are often very useful open source alternatives.

### 4.4.4 Usage

Static analysis tools comprise an important part of the quality control loop, because they are so cheap to run and rerun. Hence, in the moment the first code is written, the tools can be employed in the quality control loop and provide up-to-date information after each run in the continuous build of the software. The main challenge is the suitable configuration and adaptation of the tools to make them work efficiently in your environment. This includes the installation of the tools at the programmer's work stations and the build server, including it in the build system, connecting it to the quality dashboard, removing unsuitable checks and adding specific checks.

Installing the tools on the work stations of the programmers helps them to identify problems before they go into the source code repository. The ideal case is to have the tools integrated in the programming IDEs of the developers, so that they get immediate feedback if a tool finds a problem. In that case, you need to establish versioning of the configurations of the tools so that all developers work with the same configurations. In addition, running the tools also with the continuous build helps to double-check the rules and to aggregate and visualise the results.

If you maintain an existing code base for which you have not used static analysis tools so far, you will get an overwhelming number of warnings after their first run. It is not advisable to immediately change the code to remove all these warnings as this large change would be very costly and risky. Instead a proven approach is to run the checks and watch the trend over time. Even if you cannot remove all warnings now, you should take care not to increase the number of warnings. The existing warnings can then be removed over time in connection with other changes such as the introduction of new features.

Finally, although static analysis does not often find problems that would lead directly to a field failure, it still can help in preventing field failures. There is indication that the modules with a high number of warnings from static analysis tools are also the modules that contain faults that will lead to field failures [205,220]. Hence, you should aggregate these warnings also in the dashboard to perform a hot spot analysis of problematic modules that then should be inspected or tested more intensely.

### 4.4.5 Checklist

- Have you configured the static analysis tools so that they fit to your quality model?
- Have you discussed the requirements for the tools with your team?
- Have you adapted the used static analysis tools to further requirements of your organisation, product and project?
- Do you employ more than one tool to maximise defect detection?
- Do you use at least style checkers and bug pattern tools for all your code?
- Do you use data-flow and control-flow analysis tools for critical code?
- Have you integrated the execution and data collection of the static analysis tools into an overall quality dashboard?
- Do you watch the trend analysis of the warnings issued by the static analysis tools?

- Do you aggregate the warnings of the static analysis tools to identify problematic hot spots in your software product?
- Do you give the programmers access to the warnings directly in their IDEs if the tools allow this?

### *4.4.6 Further Readings*

- Ayewah et al. [5]
  An easy to read and general introduction to bug pattern tools with a focus on FindBugs.
- Chess and West [33]
  This book is from the people behind the tool Fortify. They describe here specifically how to use that tool to support security.
- Bessey et al. [19]
  An interesting behind-the-scene discussion of the makers of Coverity on how to build static analysis tools and also how to build a company around it.

## 4.5 Testing

With *testing*, we mean all techniques that execute the software system and compare the observed behaviour with the expected behaviour. Hence, testing is always a dynamic quality assurance technique. Testers define test cases consisting of inputs and the expected outputs to systematically analyse all requirements. We give an overview of the variety of test techniques, discuss their effectiveness and efficiency as well as how they can be automated. Using this knowledge, we describe how they are included in the quality control loop.

### *4.5.1 Regression Testing*

The basis of our approach to software quality control is to employ the control loop (Sect. 4.1) in the iterative development of the software system. Thereby, we identify defects early and soon after they were introduced. This means we also run all tests in each quality control loop to check whether the new code works and the old code is not broken. This is commonly called *regression testing*. It is a test approach in which we reapply test cases after changes. The idea is that once the test suite (i.e. a set of test cases) is developed, it can be run again after a change of the system was done (either bug fix or new feature) to make sure that existing functionality and quality has not been broken by the change. All test techniques we will discuss later on can provide test cases to be used in regression.

Already in initial development but even more so for successful maintenance, comprehensive regression testing is imperative. Making changes without regression tests is suicidal, because you can break the system without even realising it. It is not feasible, however, to do manual tests every time you commit a change to the system. Hence, regression testing must be automated as much as possible. Besides the effort and time savings, there are more reasons for this automation:

- Automated tests ensure repeatability.
- Manual tests are almost always underspecified and therefore executing the same manual tests multiple times creates very different traces in the system.

Having these automated tests, they must be run as often as possible, e.g. every night. This ensures that we catch defects early when they are still cheap and close to when they were introduced which eases debugging. Furthermore, you cannot tolerate failing tests in these nightly tests as new failures will hide in the failing tests. You will miss them, and they will then cause more trouble and cost more than necessary. Regression testing reveals failures. The causing faults, however, are often hard to find. One way to tackle this is to shorten the test cycles and run regression tests after each change, but this is not possible for large test suites. In addition, it is also the reason that regression testing should be done on all levels: unit, integration and system. On the unit or integration level, the scope of the test is smaller and this can help you in locating the fault to the failure. Having found the fault, you should use this information as a source for a new test case. You make sure that this kind of defect is definitely caught and will be easier to debug in the future.

Automated regression testing does not only require suitable test cases and a proper automation. Often it is equally hard to define a suitable test environment where, for example, the database can be reset to a defined status after each test (or collection of tests). In system testing, test execution often requires third-party systems that need to be present or mocked. While this can be set up for a singular system test session relatively easy, it is often non-trivial to create such an environment for tests that are run every night.

Unfortunately, despite its importance in modern development processes and quality control, regression testing is underrepresented in academic research. Also many textbooks discuss it only as a side aspect. I am convinced it should be the underpinning of all test efforts.

### 4.5.2 *Different Techniques*

Testing is the most important quality assurance technique in practice [212]. It is, however, not a single homogenous set of techniques, but it contains all structured executions of a software system with the aim to detect defects. The two major dimension we need to consider are the test phases and what drives the test case derivation [15, 74, 115, 161]. Figure 4.14 shows these two dimensions with concrete examples and how they can be placed according to those dimensions.

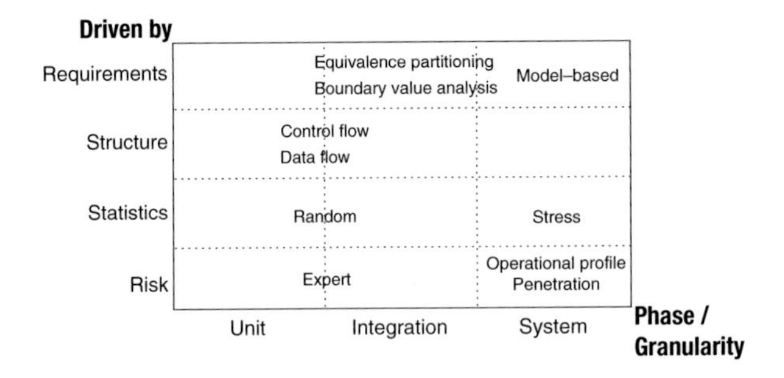


**Fig. 4.14** The two major dimensions of test techniques [202]

On the phase dimension, we have the main parts *unit test* (also called *module* or *component test*), *integration test* and *system test*. In unit tests, the testers only analyse single components of the system in isolation using stubs to simulate the environment. During integration tests, they combine the components and analyse their interaction. Finally, in system testing, the testers check the whole system either in a test environment or already the production environment. Depending on the type of system, this can include a complex software/hardware/mechanics conglomerate. In addition, if the customer makes the system test, we call it *acceptance test*. To call this dimension *phase* can be misleading, as we do not expect to run each level of tests separately one after the other in each iteration. Therefore, you can also think "granularity" here: modules, groups of modules and complete system. We will cover each phase or granularity in detail in the following:

## Phases

### Unit Test

Any modern programming language supports decomposing a system into components or units. In object-oriented languages, these are usually classes; in other languages, there are modules. These units of a software system are the smallest building blocks that can reasonably be tested. A test on the statement level does normally not provide interesting results. A unit, however, describes a piece of functionality and thereby is the interaction of several statements. The correct implementation is tested by *unit testing* against the module specification.

Unit tests are developed at the same time as the units. In many companies, the developers of the units themselves specify and perform the unit tests. In test-driven development, the developers even write the tests before the implementation and write the code to fit to the tests. Although there is empirical indication that test-driven development increases quality [32, 186], having the developers create unit tests breaks the dual control principle. This might lead to missing defects on that level. How serious this problem is in comparison to the gains, however, is not known yet.

For unit tests, automation is important, as they should be run very often. For that, open source test frameworks, such as JUnit[12] for Java, RTR[13] for C# or Tessy from hitex for embedded systems, are available for many programming languages and environments. Ideally, unit tests check every change for consequences that break an existing test. This avoids that you perform large changes that break the system in various ways, because it then becomes difficult to separate all single issues and you need a high effort to resolve all of them.

In the unit test phase, the whole "driven by" dimension can be employed and you should use different test case derivation techniques to be most effective [116]. For example, for unit tests, it is straightforward to develop white-box tests to cover the code of the unit. To find the most defects, however, you should add black-box and other tests as well.

Most important is that you concentrate on an isolated unit. This allows you to test that unit in detail, to debug easier and to test early, even if the system is not complete yet. This comes at the cost, however, that you need to build drivers to run the tests and stubs which simulate the behaviour of missing units that you use in the unit under test. Moreover, when you test the units, you will need to add additional diagnosis interfaces to the units to make them easier to test. Usually, this additional effort is outweighed by the gains of thorough unit testing.

In general, in the quality control loop you should aim at having only unit test cases in the regression that show no failures. While developing changes to the system, unit tests will break, but the developer should realise that while the change is still not checked in. The rule should be that a developer should not check in code that breaks existing unit tests. Either the code or the tests need to be changed. To ensure this, you should include a unit test run after changes with the continuous build. In case the test run is very short, you could even test with each check-in.

If unit tests shows a failure, this gives a strong indication of the overall quality of the system. It is very likely that a user could experience a failure. Hence, a test showing a failure should never be accepted. In addition to the high probability of a user-visible failure, such tests in nightly quality analyses can build up over time, and with half the tests failing, nobody will take them seriously anymore.

To be able to test isolated units, we need a software architecture that specifies such units. This does not happen automatically when we build a software, but we need to consider it explicitly in the design of the system. We need to have an architecture of testable units.

### Integration Test

While the developers create units and you test each unit by itself to assure its quality, you can start in parallel to integrate the finished units to larger subsystems. You can

[12]http://www.junit.org/

[13]http://rtr.codeplex.com

employ different integration strategies, such as bottom-up, top-down, big-bang or sandwich [171]. In all strategies, you replace stubs with real units, and in addition to the quality of the individual units, you are now interested in the quality of the interaction between units. Each unit could behave correctly to its specification and the intentions of the developers, but still their interaction could be wrong. These are interface defects. For example, consider a unit *Converter*, which converts data from another system and uses a unit *Separator* to break up a text file. *Converter* gives the text file to *Separator* and expects data chunks separated by semicolons. If the separation character was not explicitly specified, the developer of *Separator* could choose to use tabulators for separating the data chunks. Both units work correctly, but their interaction is wrong. In addition to correctness, integration testing is also interested in further properties of the interactions such as timing.

Massive parallel development of units and saving the cost for stubs seem to favour a big-bang integration. This means that you develop all units individually and then integrate them all at once. This holds the risk, however, that you get a large number of interface defects and you will struggle to sort them out. Therefore, I suggest to use incremental integration in which you add one or a small number of units after the other to build up the complete system (*incremental integration test*). This way, we can keep the number of integration problems at a manageable size. In fact, the best proven practice is continuous integration, i.e. a nightly build that integrates the current state of development and runs all unit and integration tests available so far. This combines perfectly with continuous quality control. In essence, a big-bang integration would bring you almost directly to a system test (see below) in which you continuously have to deal with simple interaction defects that still take a lot of time because the system test has to be restarted after fixing them.

Similar to unit tests, the aim should be to automate the integration tests as far as possible. Then we can include them in our continuous control loop and they help to identify interface defects in the interaction of units. In the incremental integration tests, we do not want to lose the integration we have reached so far. Therefore, we suggest to also run those integration tests that have at least once been successful in the hourly or daily build. If we employ levels of integration, from subsystem integration to system integration in an incremental integration, we have good support in localising defects.

#### System Test

After the integration test is finished, we a have a complete system. We can now test it against the specified functionality, performance and quality. We have all possibilities to mix tests driven by requirements, statistics and risks aimed at the whole system. You can start doing that in your test environment, but at some point, you should move the tests to the production environment at the customer. The final test by the customer is then the acceptance test.

Before the acceptance test, you can have user tests with actual (future) users of the system and you observe them while they interact with the system. This can have different aspects to it. For example, a user can just be given the system without any further explanations to analyse how easy it is to learn the system. Without a detailed observation, user tests can already help to identify defects in the most commonly used parts of the system.

We want to stress that for system testing, the key to find the most defects is to use different techniques to cover all the aspects you are interested in. Below we will discuss the drivers of tests in more detail which should be used by system tests. Important is that you do not only concentrate on functional correctness but that you also go for performance or security at this point.

A difficulty with system tests is that not every test case is automatable. Therefore, you cannot run all system tests continuously. This would probably exceed your available effort by far. Nevertheless, you can still automate what can efficiently be automated and plan carefully when manual tests can give you the most information about the quality of the system. In particular, system tests give important input for the evaluation of the complete functionality and performance that are not measurable in earlier phases.

### Drivers

We divide the forces that drive test case derivation into *requirements*, *structure*, *statistics* and *risk*. Requirements-driven tests only use the (functional) specification to design tests. Structure-driven testing, also called structural or white-box testing, relies on the source code and the specification. Statics-driven tests use statistical techniques to derive test cases. This can be completely random or driven by certain distributions that might also be constraint by what is sensible with respect to the interfaces and behaviour of the system. Risk-driven tests concentrate on covering the parts of the systems with the highest risks. This can be high probability of usage or failure as well as parts with the most sensitive information or the most dangerous behaviour.

#### Requirements

Requirements-driven tests only use the (functional) specification to design test cases. They are also called functional or black-box tests. Examples for functional tests are equivalence partitioning and boundary value analysis. In equivalence partitioning, the testers divide the input space into partitions that presumably trigger the same kind of functionality. Then they specify a test case for each partition. Boundary value analysis is complementary to this, as with it, the testers specify test cases for all boundary cases of the partitions.

A special case of requirements-driven testing is model-based testing where the requirements specification is substituted by an explicit and formal model. The model is significantly more abstract than the system so that it can be analysed more easily (e.g. by model checking). The model is then used to generate the test cases so that the whole functionality is covered. For example, if the model is a state chart, state and transition coverage can be measured. If the model is rich enough, the test case generation and even the expected output can be generated automatically from the model. The model then also acts as oracle, i.e. it knows the correct output to an input. These benefits stand against the effort to build and maintain the model. With frequent requirements changes, it can be very efficient, because only the model needs to be changed but the test cases can be generated.

To drive tests by requirements includes the need that there are clear and complete requirements specifications. If they are not available, the tests themselves become the first formalisation of a requirement beyond code in the project. In small projects, this can be a valid and useful way. In larger projects, however, it is a very late point in the project to concretise requirements and it could lead to a lot of additional, unnecessary rework.

Requirements-driven tests are the basic part of tests in the quality control loop. We specify the requirements, functional requirements and quality requirements, from the quality model and check the conformance of the system with them. You should start as early as possible to formulate those requirements and reuse quality requirements and their operationalisation from other projects to save effort.

#### Structure

Structure-driven testing, also called structural, white-box or glass-box testing, relies on the source code and the specification. We divide structural testing into *control-flow* and *data-flow* techniques. In control-flow testing, the testers specify test cases according to the different possible paths of execution through the program. In contrast, they define test cases following the reading and writing of data in data-flow testing. There are measures to analyse the completeness of both kinds of structural tests. A set of test cases, a test suite, covers the control or data flow to a certain degree. For control-flow coverage, we measure the share of executed statements or conditions, for example. For data flow we measure the number and types of uses of variables.

Moreover, modern software systems come with a huge amount of parameters that can be set to configure the behaviour of the system. These settings are also called configurations. In practice, it is usually impossible to test all configurations because of the combinatorial complexity. Configuration testing is the method that generates test cases out of possible configurations in a structured way.

It is important to note that while structure-driven tests are important and necessary, because they give you a different point of view than requirements-driven tests, you should not overemphasise them. It is easy to measure a certain coverage of code by those tests. It is unclear, however, what that means for the quality of your system.

With the coverage by structure-driven tests, you only know that you have at least executed a statement or branch while testing. This is a basic indicator, but it is hard to interpret. If you only have a low coverage, it indicates bad quality because you missed many parts of your system with your tests. If you have a high coverage, you still can have bad quality because there can be system parts missing or the covered parts do not follow the requirements correctly. Hence, you need to use more than just structure-driven tests, but they give you a basic indication of quality. Furthermore, in case your software system needs to conform to safety standards such as IEC 61508 or DO 178B, you are required to perform and document code coverage.

#### Statistics

Statistics-driven tests use statistical techniques to derive test cases. This can be completely random or driven by certain distributions that might also be constrained by what is sensible with respect to the interfaces and the behaviour of the system.

Random tests are generated without the specific guidance of a test case specification but use a stochastic approach to cover the input range. It does not rely on the specification or the internals of the system. It takes the interface of the system (or the unit if it is applied for unit testing) and assigns random values to the input parameters of the interface. The tester then observes whether the system crashes or hangs or what kind of output is generated.

Performance/load/stress tests check the behaviour of the system under heavy load conditions. They all put a certain load of data, interactions and/or users on the system by its test cases. In performance testing, usually the normal case is analysed, i.e. how long does a certain task take to execute under normal conditions. In load testing, we test similar to performance testing but with higher loads, i.e. a large volume of data in the database. Stress testing has the aim to stretch the boundaries of what is possible or expected load on the system. The system should be placed under stress with too many users interactions or too much data.

#### Risk

In risk-based testing, test cases are derived based on what in the system has a high risk. The methods assess risks in three areas:

- *Business or Operational.* High use of a subsystem, function or feature, criticality of a subsystem, function or feature, including the cost of failure
- *Technical.* Geographic distribution of development team, complexity of a subsystem or function
- *External.* Sponsor or executive preference, regulatory requirements

All of these areas can trigger specific test cases. For the business or operational risks, we employ, for example, testing based on operational profiles. They describe how the system is used in real operation. This usage then drives which test cases are executed. For example, imagine *manager* is a component in our system. Its subcomponents *input* and *output* are used differently. In 30 % of the usages, the *input* component, in 70 % the *output* component is executed. Our test suite follows this with 70 % test cases that invoke *output* and 30 % test cases that call *input*. This way, tests driven by operational profiles act as a good representation of the actual usage.

The most common risk-based testing is probably *error guessing*, which belongs to the technical area. Experts familiar with the system and the domain predict which parts of the system and which situations and use cases are most likely to contain defects. You can then build test cases to cover these risks.

Penetration testing is an example for a test technique that considers criticality of a function. It derives test cases from misuse cases. Misuse cases describe unwanted or even malicious use of the system. These misuses are the basis for penetrating the system to find vulnerabilities in it. The result can be to either

- Crash the system,
- Get access to non-public data or
- Change data.

### 4.5.3 Effectiveness and Efficiency

Overall, testing is an effective quality assurance technique. On average, it is able to detect about half of the defects [200]. If you perform tests thoroughly and skip inspection before the tests, this value can go up to 89 %. If you perform tests poorly, however, the effectiveness can drop to 7 %. These values are similar over all major testing techniques.

One reason for the popularity of testing is probably that it is effective in many defect types. It is the most effective for data, interface and cosmetic defects [11]. Data defects are incorrect uses of data structures, interface defects are incorrect usage or assumptions on the interface of a component and cosmetic defects are spelling mistakes in error messages, for example. Structural tests are, in addition, effective for initialisation defects and control-flow defects. For many of the more specialised test techniques, there is no conclusive empirical knowledge.

The effort for finding a defect is also very similar over all test techniques. On average you need to spend 1.3 person-hours to find a defect with a range from 0.04 to 2.5. A defect, however, needs also to be removed. The test phase dimension differentiate here more clearly. It becomes the more expensive to remove a defect, the later it is removed [23, 200]. In unit testing, it takes 3.5 person-hours to remove a defect, in integration testing already 5.4 person-hours and in system testing 8.4 person-hours. These values are all higher than for automated static analysis or reviews and inspections.

### 4.5.4 Automation

For fast feedback to the developers in the quality control loop, we need to be able to run tests frequently. Manual tests are too expensive to execute on a daily basis.

Therefore, it is necessary to automate at least part of the tests so that they can be run often. Three parts of the test process can be automated: (1) generation of the test cases, test data and parameters; (2) execution of the test cases; and (3) evaluation of the test results in comparison with the expected results.

#### Generation

The generation of suitable test cases is intellectually the most difficult part in testing and therefore it is hard to automate. In practice, experienced testers and experts of the domain and system look for difficult and critical scenarios that are beneficial to test. In addition, they aim to cover the software – its code and requirements – with their tests so that they do not forget to test parts of the system. The goal is to test as complete as possible.

Especially these coverage criteria are the starting point of test case and test data generation. It is possible to generate test cases completely randomly, but this results in many invalid and useless test cases. In contrast, we can employ the code or models of the requirements to guide the test case generation. The stop criteria for the generation can be coverage of the code or the models.

In practice, generation of test cases is interesting, because a lot of effort is put into doing this manually. The effort that has to be spent for the automation, however, is also substantial. In our experience, it is most useful in situation in which vast amounts of test data or very long sequences of interactions are necessary to test a system. For example, a database-intensive business information system needs many business data or a software that implements a network protocol needs to be brought to its boundaries by creating long sequences of inputs and outputs.

#### Execution

In many software projects, the testers only automate test case execution. The challenges here lie in creating a test environment (system under test, stubs, hardware) in which tests can run completely automated as well as in maintaining the automated test cases. For certain programming languages and technologies, there is mature tool support for creating executing tests. For example, unit tests in Java can build on the JUnit framework which encapsulates many issues around executing and evaluating test cases. But also with a good framework, executable test suites can become very large and they need to be adapted to changes in the system. Therefore, they are a development artefact similar to the code that it tests. Hence, the same level of care and quality assurance is necessary for them. Then automating the test execution has a high potential to pay off.

#### Evaluation

Evaluating tests means that we compare what we got as a result from the tested system with the expected result. Depending on the type of system and output, we can tolerate small deviations. For example, in real-time systems, we need to check whether messages arrive in a certain time frame. The expected results that we compare the actual results with can come from two sources: (1) the tester derived them from the specification or (2) an oracle. In many cases, while making the test cases executable, it is straightforward to implement the evaluation along with the test case execution. In some cases, however, calculating the correct result is difficult. For example, if the software under test does complex computations of aerodynamics, the tester cannot calculate that result by hand. Then an additional automated mean is needed – an oracle. For example, there might be simulation models of these aerodynamics which also produce a result that can be employed for the evaluation automation. If you used executable models for test case generation, they are also able to generate expected results.

### 4.5.5 Usage

Testing is essential in the quality control loop as it can directly show how a software system fails. If you work with our Quamoco quality models (Sect. 2.4), all activities in the model that involve end users need to be analysed by testing in addition to other quality assurance techniques, because testing can best resemble the actual user experience. Nevertheless, also testing has a limited defect detection capability and it tends to be more expensive than other techniques. Therefore, it should not be the sole method that you use in quality control. Only a combination can give you the optimal relation between effort and found defects.

The variety of testing techniques is large and each technique has different advantages and disadvantages. It also applies here that a combination is most advisable. What test techniques you should employ depends on the quality needs of your software system. As most experts consider functional suitability as the most important quality attribute, you should use functional tests that cover the specified requirements and additional expected properties. To improve defect detection, you can combine them with structural tests and measure code coverage.

For any non-trivial system, it is advisable to employ (automated) regression testing. The investment in these tests quickly amortises, because you save hours of debugging failures after changes to the system. Especially if you have regression tests on all levels of granularity, fault localising will be far cheaper. If your regression test suite becomes too large to be run completely after each change, you should at least run it in the nightly build and you can choose the most important test cases for each change.

If you need to use other test techniques depends largely on the type of system and its quality needs. In high-reliability or security environments, random testing is a good candidate to check the robustness. Reliability growth tests and penetration tests help in assuring these characteristics in a more focused way. In many cases, especially if there is a high risk, we propose to use performance tests.

### 4.5.6 Checklist

- Do you plan for and have an explicit architecture consisting of testable units?
- Have you planned for explicit unit, integration and system tests?
- If you plan to skip unit or integration tests, do you have a good reason for it?
- Do you employ a mixture of different test derivation techniques to maximise defect detection?
- Do you use other quality assurance techniques in combination with testing?
- Have you automated tests as far as possible to enable regression testing?

### 4.5.7 Further Readings

- Myers [161]
  The standard book on software testing. It has grown old a little, but it still contains the basics you need to know, described in an easy-to-read way.
- Beizer [15]
  A long-recognised book on software testing. It explains the whole process and many different techniques. It contains an especially detailed classification of defects.
- Spilner et al. [193]
  A book that focuses on what we left out in this section: test management. It gives a good overview and prepares you for the certified tester exam.
- Graham and Fewster [78]
  Although over 10 years old, this book still gives good insights into test automation.
- Broy et al. [28]
  One of the early comprehensive books on the available research in model-based testing.
- Beck [14]
  If you are interested in the idea of test-driven development and you want to learn more about how you can use it, this book is a good start. It is from one of the main founders of TDD and it is well written.

## 4.6 Product and Process Improvement

If applied successfully, continuous quality control allows to counter quality decay. Beyond that, it also provides additional insights that enable an organisation to *improve* quality and its *quality management* practice. This chapter discusses how the insights gained during product quality control can be used as input for improvements in the quality of processes, the quality models, the skills of the team and the overall product. Hence, product quality control feeds information back into the overall continuous improvement for the PDCA cycle (see Sect. 4.1).

*Quality control* ensures that a product meets the defined quality requirements as deviations are detected and corrected. It does not automatically provide *quality improvement*, however; i.e. it does not raise an organisation's ability to produce high-quality products (the *manufacturing view* of Sect. 1.3). To provide this, you should translate the insights gained during quality control to incremental improvements of the development and quality assurance processes.

### 4.6.1 Lean Development

Process improvement comes down to changing the way people work. The work flows we practice in software development, however, are far too complex and variable to completely document them in a process description and then simply change that documentation to improve the process. Instead, we need the right mindset embraced by all employees. Everyone who participates in the development process needs to strive for excellence and thereby works for a high-quality product.

Such a mindset is what is behind many well-known approaches such as Total Quality Management (TQM), Six Sigma or ISO 9000. We found, however, that a lean development approach is the easiest to implement in practice and beneficial in most contexts. Especially the notion of *wastes* (originally *muda*) in the process characterises exactly what we look for in process improvement. Wastes are unnecessary steps in your process that do not add to the value (or the quality) of the product. In the Toyota Production System, from which lean development grew, we find these classes of wastes [154, 218]:

- Overproduction
- Waiting
- Handoff
- Relearning/reinvention
- Partially done work
- Task switching
- Defects
- Under-realising people's skills
- Knowledge loss
- Wishful thinking

All of them stem from problems they found at the Toyota production lines, but we can easily transfer them to software development [179]. Are we adding features the users do not need? Are we building components that are already available in libraries? Do we document our design decisions? All these questions lead to potential wastes in your process and all members of the development team need to feel responsible for looking for these wastes and to get rid of them.

Wastes are a vivid notion for what we should avoid. They are, however, only a part of lean development. The two central principles are *continuous improvement* and *respect for people*. We should all the time look for possibilities to advance our processes in a way that supports the people involved in these processes.

Continuous improvement includes recognising and removing wastes, but it also contains regular, so-called Kaizen events in which you identify small, incremental changes to the process. Implementing only these small changes improves the probability that you can succeed in changing the process, because modest changes in behaviour are far easier to make permanent. A single but large process change is likely to be refused by the team. The Kaizen event is a dedicated workshop with your team or a subset of your team to identify process improvements. It does not only consist of identifying wastes, but you can also decide to introduce additional process steps, for example, to increase cooperation with other teams or to improve quality standards. Holding it as a team meeting instead of letting the team leader decide also helps that the team will accept the process changes. In addition, continuous improvement denotes by *continuous* that there is no final process. There is always something to improve.

The respect for people constitutes an even more important part of process improvement in software development. Building or maintaining software is a task that extremely depends on the people that carry it out. Only highly motivated and supported team members can build high-quality software. This motivation and support need the corresponding respect.

How does that fit into continuous quality control and the control loop? On a higher level of abstraction, the mindset of lean development should give you a good basis for steering a project and developing your team. On a lower level, lean development's continuous improvement depends on the recognition of wastes that can be discussed at Kaizen events. The team members might recognise wastes while they perform the current process. But also the analysis of product quality in the quality control loop can give indications of wastes.

### 4.6.2 Derivation of Improvement Steps

In the quality control loop, the quality engineer gets feedback from the quality analysis on the current state of the product's quality. In this context, we need to distinguish internal quality analyses, which are done in-house, and external quality analyses, which collect field failures and customer satisfaction. Both kinds of analyses indicate that we need a better constructive quality assurance that prevents these quality defects. Results from external analyses, however, show in addition that we need to add more analytical quality assurance – internal quality analyses – so that the quality defects do not make it to the users.

In any case, for each quality defect reported from quality analyses, we need to analyse the root cause. Instead of only trying to get rid of the symptoms, we need to

**Problem:** Our software performs badly at a specific customer.

1. Why? Users wait minutes for queries to give results.
2. Why? The object-relational mapper needs too long for the mapping.
3. Why? The database system used by the customer was not tested.
4. Why? We concentrated on a standard configuration without communicating the compatibility list to our marketing.
5. Why? There is no interface between development and marketing that clarifies compatibility requirements.

**Solution:** Install explicit role in the process who is responsible for clarifying compatibility requirements.

**Fig. 4.15** An example of the 5 Whys method

find and remove what caused the defect. In many cases, this is trivial. A developer misunderstood the requirements or forgot to implement a special case. For other problems, the root cause is very hard to find. Why does the software perform badly at a specific customer site?

A simple method from lean development can help here to get to the bottom of the problem. The 5 *Whys* want you to do what the name says: ask several times *why?* to understand cause and effects of a problem. Five is not a fixed number of questions but is a rule of thumb born out of experience. You can, however, also ask *why?* six or seven times. Let us look at the example in Fig. 4.15.

Having the root cause, the derivation of an improvement still tends to be complicated. If a developer misunderstood the requirements, the possibilities for improvement are endless. We could, for example, add more requirements engineers to the team so that they write more detailed requirements specifications. Or we change the way we specify requirements and add UML sequence diagrams and state machines. Which improvement fits best to your root cause depends on your context and we cannot give a clear guidance on that.

Nevertheless, it is a best practice to find small improvements that are incremental to the current process. You can, for example, find and discuss them in your regular Kaizen events. Revolutionary changes hold a high risk and should only be implemented in rare, extreme cases. Furthermore, it is desirable to have a measurable improvement so that we know when we have finished the improvement. You can feed this measurement back to the quality control process and thereby control when you reached your improvement goal.

### 4.6.3 Implementation of Improvement Steps

Implementing process improvements means changing the behaviour of people. This is only possible, if these people are also convinced of this change. How to implement change in an organisation has established itself as an area of research in its own right [129, 130, 133]. We take a simple five-step method based on [181] from that research that helps you in implementing change. The five steps are:

- Convince stakeholders
- Create a guiding coalition
- Communicate change
- Establish short-term wins
- Make change permanent

### Convince Stakeholders

We start by making a list of all stakeholders that are affected by an improvement measure. These stakeholders need to be convinced that the improvement is necessary and effective, even if there is a spirit of continuous improvement. We need the support of all people that are involved to make it a success.

We can achieve that by showing dramatic or surprisingly new information and data about quality defects, their effects on the users and the company, as well as the reasons for these quality defects. It is important that you touch the stakeholders on an emotional level to motivate them for change. Also keep in mind that different stakeholders will have different motives that you need to consider in convincing them. The test team might be interested in reducing certain defects so that they have less to report, the development team so that they have less rework.

### Create a Guiding Coalition

It is easier to establish a process change, if not only one manager is responsible for it, but if there is a team that is the guiding coalition. This team should include people with certain characteristics: enough power to make decisions during the change implementation, broad expertise to make sensible decisions, credibility to explain the change, management to control the change and leadership to drive the change.

### Communicate Change

The guiding coalition shows the goal behind the improvement to group leaders and developers. It can be frustrating for team members to change their routines without seeing what it is good for.

### Establish Short-Term Wins

The best way to further motivate the people for keeping the process improvement is by giving them visible, measurable short-term wins. For example, by introducing a static analysis tool, which the developers need to run before they check in code, we can build diagrams that show that it then reduces the defects found in inspections and unit tests and thereby the rework that the developers need to do.

Make Change Permanent

Short-term wins, however, do not realise the full potential of a process improvement. Therefore, we need to incorporate more and more employees into the change process. Furthermore, it helps to promote successful changes and to appoint new employees that help in using the new process. For example, a new employee could be responsible for selecting suitable static analysis tools, to keep them up to date and to integrate them into the standard development environment.

In the end, after you ensured that the change is permanent, you are left with the responsibility to check whether the improvement actually made something better. The control loop can help you to track whether the problems that brought you the process change persist or come up again. In that case, you need start from the beginning and ask more why questions to find the real root cause. If the problems are gone, congratulations! But remember that process improvement is continuous, because there is no perfect process.

### *4.6.4 Checklist*

- Have you established continuous improvement and respect for people as key principles in your organisation?
- Do you check each quality analysis result for root causes and potential process improvements?
- Do you have regular Kaizen events to discuss wastes in your process and how you could remove them?
- Have you convinced all stakeholders of the process improvement?
- Have you created a guiding coalition?
- Have you communicated the change to all employees?
- Have you established short-term wins?
- Have you made the change permanent?

### *4.6.5 Further Readings*

- Monden [154]
  The book on the initial Toyota production system. It is not directed at software development, but it gives you the historical background to lean development.

- Womack et al. [218]
  The story from Henry Ford to lean development written from an American perspective.
- Poppendieck and Poppendieck [179]
  Probably the earliest comprehensive application of the lean principles to software development.
- McFeeley [150]
  A handbook on software process improvement from the SEI. It consists of phases that you can map to our steps. These phases are, however, more deeply described.

# Chapter 5
# Practical Experiences

This chapter describes several practical experiences we have made over the last 10 years with different parts of the product quality control approach described in this book. The first three experience reports concentrate on building quality models and using them for quality requirements and evaluating quality: the Quamoco base model, the maintainability model for MAN Truck and Bus, the security model for Capgemini TS and an activity-based quality model for a telecommunications company. Next, we describe the application of quality prediction models, in particular reliability growth models, at Siemens COM. Finally, in the last experience report, we focus on applying analysis techniques: We apply static analysis, architecture conformance analysis and clone detection at SMEs.

## 5.1 Quamoco Base Model

The Quamoco base model is the largest and most comprehensive application of the Quamoco approach today. Therefore, we start with the experiences we have made in building it and applying it to several open source and industrial systems.

### *5.1.1 Context*

Quamoco was a German research project sponsored by the German Federal Ministry of Education and Research[1] from 2009 to 2013. You can find all information on the project also on its website http://www.quamoco.de. The project was conducted by a consortium of industry and academia, all with prior experiences with quality models and quality evaluation. The industrial partners were Capgemini, itestra,

[1] Under grant number 01IS08023

S. Wagner, *Software Product Quality Control*, DOI 10.1007/978-3-642-38571-1_5, 

SAP and Siemens. The academic partners were Fraunhofer IESE and Technische Universität München. JKU Linz participated as subcontractor for Siemens.

As this was a large project with many partners, there were different motivations and goals in joining the consortium. Therefore, we created various different contributions in the project. The main aim of all members of the consortium, however, was to build a generally applicable model that covers (a large share of) the main quality factors for software. Over time, we decided to call this the *base model*, because in parallel we were working on several specific quality models targeted at the domains of the industrial partners. For example, Siemens built a quality model for software in embedded systems [146].

What is also remarkable about Quamoco is the strong commitment to tool development. We built a quality model editor that supports in building such comprehensive quality models including all additional descriptions and quality evaluation specifications. It allows different views on the model to aid understanding, and it even has a one-click solution to perform the evaluation of a software system for the base model. This is connected to the integration with ConQAT (see Sect. 4.1) which executes all analysis tools, collects the results for the analyses and calculates the evaluations using the evaluation specifications from the model. Although the user-friendliness of the tool chain could be improved, the overall integration of model building and quality evaluation is probably unique.

### 5.1.2 Building the Model

The base model should be a widely applicable and general quality model. Therefore, we could not directly follow the model building approach of Sect. 3.1. The approach is more suitable for building specific quality models. Nevertheless, the main steps are still there but with rather broad results. The relevant stakeholders of the base model are all common stakeholders in software engineering: users, customers, operators, developers and testers.

Hence, the general goal is to build high-quality software. We referred to ISO/IEC 25010 (Sect. 2.3) as relevant document to elicit our quality goals which were the quality characteristics of the standard quality model. They formed our quality aspect hierarchy in the Quamoco base model. We chose these characteristics despite the problems we discussed, such as overlap, because we assume that using the standard will result in a broad acceptance.

We experimented with various artefacts created in software development but decided to concentrate on the main artefact developers work with and which has a strong effect on the user: the source code. Hence, the base model is source-code-focused. To be able to finish the model in the project duration, we decided, driven by the industry partners, to concentrate on Java and C# for modelling the measurements. With more time, however, we would have liked to include other artefacts to better reflect their influence on the quality aspects as well as other programming languages to make the base model more widely usable. With the

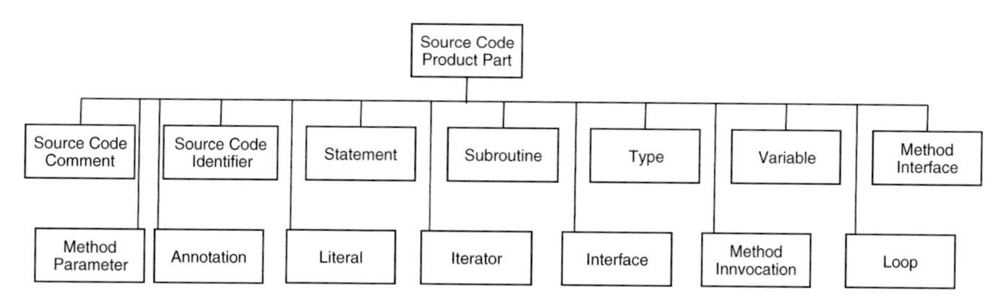


**Fig. 5.1** An excerpt of the entity tree of the Quamoco base model

decision for source code, the resulting entities are almost all source code parts. An exception is, for example, the entity identifier which is also possible in other kinds of artefacts. We created a hierarchy for the source code parts which was partly built *bottom-up* while adding product factors and measures and partly structured *top-down* to keep an understandable entity tree. Figure 5.1 shows the layer of entities below the source code part which contains most entities. There would be some refactorings possible, for example, to organise the method-related entities in a part-of hierarchy, but we found that this does not improve readability.

To come up with appropriate product factors, we drew mainly from the descriptions of analysis rules of static analysis tools we employed, the experiences we have made in other contexts and our own expert opinion. This resulted in a wide range of product factors including [Class Name | CONFORMITY TO NAMING CONVENTIONS], [Duplication | SOURCE CODE PART] and [Class | SYNCHRONIZATION OVERHEAD].

The project had the length of 3 years, and we decided early on to work in three iterations. Hence, we rebuilt and fundamentally restructured the whole model two times. We worked on specific aspects in a large number of workshop sessions with a varying number of participants from most of the consortium's members. We had many and long heated discussions about identifying and naming product factors as well as the influences of them onto quality aspects. We believe these long discussion resulted in a highly readable and understandable model.

As the base model is a general model, we could not formulate concrete quality requirements but had to model what is "normal" and generally applicable. Therefore, we searched for ways to accomplish that. For the weights of the quality aspects, which usually are specific for a product, we chose to use the results of a survey we have also done in the context of Quamoco [213]. It represents the average opinion of over hundred experts. Therefore, we modelled the weights corresponding to the importance of the quality aspects given by the survey respondents.

Finally, to determine the evaluation specifications for the product factors, we used a calibration approach based on open source systems. The assumption was that by using a large number of existing systems, we will cover many different contexts and, on average, get a widely applicable evaluation. We chose for the C# part 23 systems and selected over hundred systems with varying sizes from an open source collection of Java systems. We needed to determine the *min* and *max* for the evaluation specifications (Sect. 4.2) which we chose as the 25 % and 75 % percentiles of all the results we got from measuring the open source systems.

### 5.1.3 Quality Evaluation and Measurement

It was important for the Quamoco consortium not only to build abstract quality aspects and some corresponding product factors but to have it completely operationalised. To make this possible, we had to concentrate on two programming languages: Java and C#. For these two languages, there are measures for all product factors and corresponding evaluation specifications up to the quality aspects.

The process of measurement and quality evaluation is well supported by the Quamoco tool chain consisting of the quality model editor and an integration with the dashboard tool ConQAT. We can add evaluation specifications to any product factor and quality aspect and define how it is evaluated depending on measurement results or influencing product factors. For that, we can define a maximum number of *points* a product factor can get, distribute the points among the related measures to express weights and finally define linear increasing or linear decreasing functions with two thresholds. The actual number of points, which is then normalised to a value between 0 and 1, is determined by running the evaluation specifications in ConQAT. The model editor allows to export a corresponding specification for it.

Let us look at the example of cloning (Sect. 4.4.2) and how it is represented in the base model. We chose to call the corresponding product factor [Source Code Part | DUPLICATION]. It is not directly language-specific but could be valid for any kind of system or programming language. The model defines it as "Source code is duplicated if it is syntactically or semantically identical with the source code already existing elsewhere". We cannot measure the semantic identity directly. It might be possible to find semantic clones in reviews, but it is not very reliable. Instead, we focus on the syntactic redundancy.

Figure 5.2 shows the screen from the quality model editor in which we specify the evaluation of the product factor. We have two measures available in the model: clone coverage and cloning overhead. As their meaning is similar, we chose only to include clone coverage to measure duplication. Therefore, the maximum defined number of points (here: 100) goes to clone coverage (CP in the table). Clone coverage was the probability to find a duplicate for any randomly chosen statement. Hence, the larger the clone coverage, the higher the degree of duplication. Therefore, we chose a linear increasing function.

This leaves us with deciding for the *min* and *max* thresholds. In this case, below the *min* threshold, the product factor receives 0 points, and above the *max* threshold, it receives 100 points. Based on the calibration for Java, we found that most systems have a clone coverage between 0.0405 and 0.5338. This was consistent also with our observations. Hence, we used them as thresholds for the linear increasing function.

Let us look at a concrete example of applying the evaluation to a real system. The Java open source library *log4j*[2] is often used to implement logging. We analysed version 1.2.15 using our base model. ConQAT gave us a clone coverage result

[2]http://logging.apache.org

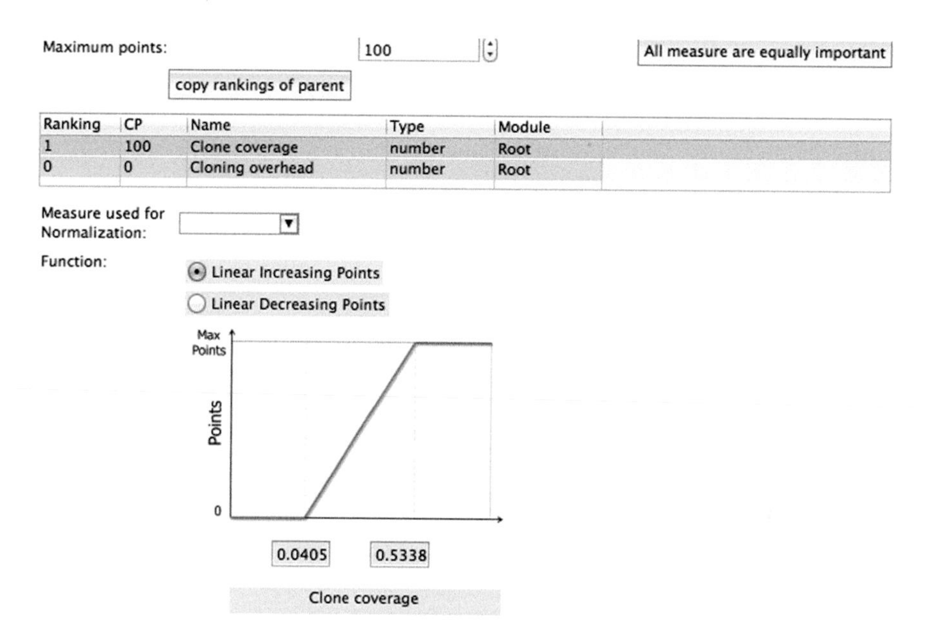


**Fig. 5.2** The evaluation specification for [Source Code Part | DUPLICATION] in the quality model editor

of 0.086 which is comparably small but above the threshold. Putting this clone coverage in the linear increasing evaluation function resulted in 9.2 points or, normalised between 0 and 1, an evaluation of the product factor of 0.092. This is very low and has a positive impact on analysability and modifiability.

Analysability, for example, is influenced by more than 80 product factors. They are weighted based on an expert-based ranking which results in a weight of only 1.06 of 100 for our product factor. For the example of log4j, although the result for the duplication was very good, the other product factors brought it down to an evaluation of 0.93 which we normalise and transform into a school grade of 4.3 (where 1 is best and 6 is worst).

As we use ConQAT for evaluating based on the base model, we directly get a dashboard created. Figure 5.3 shows an example output of the dashboard. It recreates the hierarchies of quality aspects and product factors and allows to drill down to each individual measure. In addition, it shows configuration and execution time information of the evaluation.

### *5.1.4 Applications and Effects*

Inside the Quamoco project, we applied the Quamoco quality evaluation based on the base model to several open source and industrial systems to investigate the

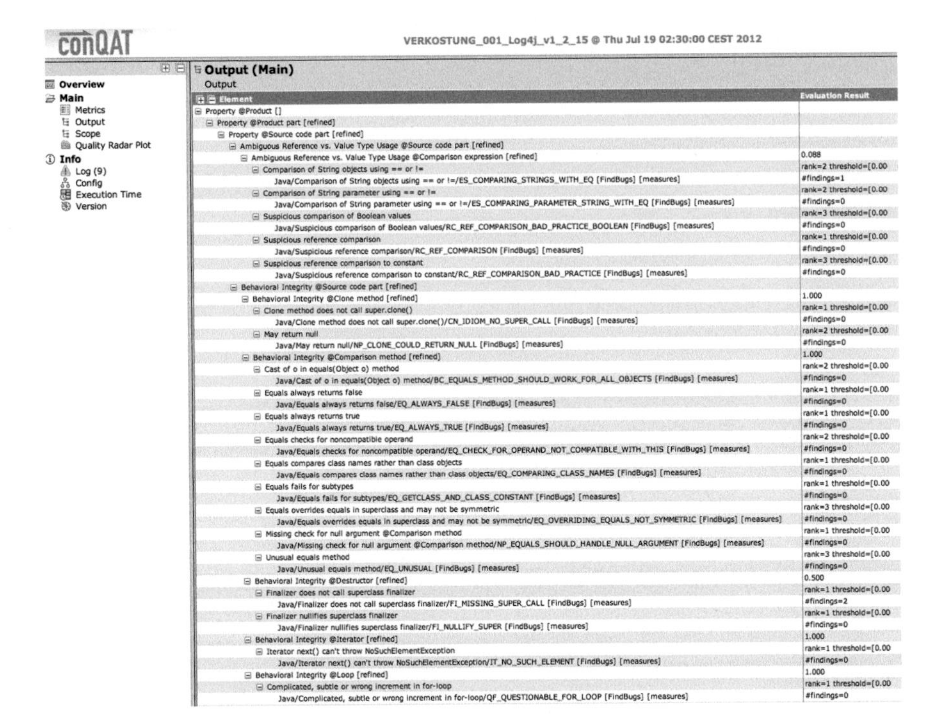


**Fig. 5.3** Dashboard for the quality evaluation

**Table 5.1** Comparison of the evaluations using the base model and the expert evaluations for five open source products [211]

| Product | LOC | Quamoco grade | Quamoco rank | Expert rank |
|---|---|---|---|---|
| Checkstyle | 57,213 | 1 | 1 | 1 |
| log4j | 30,676 | 3 | 2 | 2 |
| RSSOwl | 82,258 | 3 | 2 | 3 |
| TV-Browser | 125,874 | 4 | 4 | 4 |
| JabRef | 96,749 | 5 | 5 | 5 |

validity of the evaluations as well as how well practitioners understand the quality model and the evaluation results.

To be able to investigate the validity of the quality evaluation, we need software systems with an existing, independent evaluation. Our first application was to the five open source Java products *Checkstyle*, *RSSOwl*, *log4j*, *TV-Browser* and *JabRef*. For these products, there have been an event where experts evaluated their quality [79]. We then applied our Java base model to the same versions and compared the resulting quality ranking from our Quamoco evaluation and the evaluation from experts. You can find the comparison in Table 5.1. The Quamoco evaluation produced almost exactly the same ranking as the experts did. We were only not able to distinguish between RSSOwl and log4j.

**Table 5.2** Comparison of the evaluations using the base model and expert evaluations for five subsystems of an industrial product [211]

| Subsystem | Quamoco rank Quality | Expert rank Quality | Quamoco rank Maintainability | Expert rank Maintainability |
|---|---|---|---|---|
| Subsystem D | 1 | 1 | 1 | 1 |
| Subsystem C | 2 | 4 | 3 | 4 |
| Subsystem B | 3 | 2 | 5 | 3 |
| Subsystem E | 4 | 4 | 4 | 3 |
| Subsystem A | 5 | 2 | 2 | 2 |

**Table 5.3** Quamoco evaluation results for consecutive versions with explicit quality improvements [211]

| Version | Quamoco grade |
|---|---|
| 2.0.1 | 3.63 |
| 2.0.2 | 3.42 |
| 2.1.0 | 3.27 |
| 2.2.1 | 3.17 |

Next, we looked at an industrial software and an expert ranking for its five subsystems. We used that again to compare it to the corresponding Quamoco evaluation results. You can see the comparison for quality overall as well as maintainability, which is the strongest part of the base model, in Table 5.2. The agreement in the rankings is not as clear as above, but the general trends are similar. There is an agreement in the best ranked subsystem D. Subsystem A, however, came out last in overall quality, while the expert saw it much better. One explanation could be that the expert put more emphasis on the maintainability of the product in which the Quamoco evaluation also judged subsystem A as better. Overall, this application also supported our choice of contents and calibration of the base model.

A third application of the base model was to another industrial system that underwent quality improvements specifically in some releases. We were interested in the question whether the Quamoco evaluation results would be able to show a quality improvement. Table 5.3 shows the small but gradual improvement as evaluated by the Quamoco base model. Hence, at least for this case, the base model is well enough calibrated to support project managers in quantitatively controlling quality improvements.

Our final application was to eight industrial systems in which we evaluated the systems and afterwards interviewed experts on the systems. You can find details on the questionnaire and results in [210], but we will only highlight some of the results. The experts judged our base model to be a more transparent definition of quality than only ISO/IEC 25010, and it also could be more easily adopted and applied in practice. They also found the relationships in the quality model acceptable and the calculations in the evaluation transparent. It was seen as problematic to understand the rankings we used for determining weights in the evaluation specifications and the calibration thresholds. It was seen positively that the evaluation used school grades and that it clarifies metrics as well as the abstract quality characteristics of ISO/IEC 25010. Overall, we received very positive feedback which supports our confidence that the Quamoco base model is ready to be applied in practice.

### *5.1.5 Lessons Learned*

There were several problems in building and applying the Quamoco base model. First, we learned that building a fully operationalised quality model means that we had to build a really large model. As any other artefact, the quality model then becomes hard to maintain, for example, when we wanted to add new tools and measures which then had the effect that we needed to restructure the product factors. With the help of the quality model editor and some practice, this was manageable, but it should not be underestimated.

Second, it was hard to build such a detailed model in consensus with a group of different people from different domains. We spent many hours discussing various aspects of the meta-model, the structuring of the quality model, the impacts and even names of entities, product factors and measures. In the end, however, I believe that this has led to the good feedback we received from practitioners that they were able to understand the chosen relationships and that it helped in clarifying ISO/IEC 25010. It seems like it is a necessary process.

Third, the large amount of work we spent in building the operationalised quality model also payed back in the sense that we actually have now an almost fully automatic way to evaluate software quality for Java and C#. The model is not well equipped for quality factors that have a strong dynamic side, like performance efficiency or reliability, but even for those, we could reach reasonable results. In addition, we have a set of manual measures that need to be collected in reviews. It seems these manual measures are important for a good evaluation, but this is subject to further research.

## 5.2 MAN Truck and Bus

MAN Truck and Bus was one of the first applications of the activity-based quality model approach. Because of continuing development, the model we built there had been the most refined quality models before the start of Quamoco. The model concentrates on describing maintainability with a focus on embedded, automotive systems modelled in Matlab Simulink and Stateflow that have the aim to generate code directly from the functional models. The original description of this model can be found in [51].

### 5.2.1 Context

MAN Truck and Bus is a German-based international supplier of commercial vehicles and transport systems, mainly trucks and busses. It has over 30,000 employees worldwide of which 150 work on electronics and software development.

The organisation brought its development process to a high level of maturity by investing enough effort to redesign it according to best practices and safety-critical system standards. The driving force behind this redesign was constantly focusing on how each activity contributes to global reliability and effectiveness. Most parts of the process are supported by an integrated data backbone developed on the eASEE framework from Vector Consulting GmbH. On top of this backbone, they have established a complete model-based development approach using the tool chain of Matlab/Simulink and Stateflow as modelling and simulation environment and TargetLink of dSpace as C-code generator.

Matlab/Simulink is a model-based development suite aiming at the embedded systems domain. It is commonly used in the automotive area. The original *Simulink* has its focus on continuous control engineering. Its counterpart *Stateflow* is a dialect of statecharts that is used to model the event-driven parts of a system. The Simulink environment already allows to simulate the model to validate it.

In conjunction with code generators such as Embedded Coder from MathWorks or TargetLink by dSpace, it enables the complete and automatic transformation of models to executable code. This is a slightly different flavour of model-based development than the MDA approach proposed by the OMG.[3] There is no explicit need to have different types of models on different levels, and the modelling language is not UML. Nevertheless, many characteristics are similar and quality-related results could easily be transferred to an MDA setting.

### 5.2.2 Building the Model

Here we follow the approach for model building introduced in Sect. 3.1.

#### Define General Goals

MAN Truck and Bus, especially the stakeholder *management*, has the general goal of a global development with high reliability and effectiveness. The other relevant stakeholders in this case are the *developers* and *testers*. Both are interested in an efficient and effective development and test of their systems.

[3]http://www.omg.org/mda/

### Analyse Documents Relevant for General Goals

To further understand and refine these general goals, we analysed the existing development process definition, guidelines and checklists, as well as slide decks describing the strategy. In addition to the document analysis, the major source of input was in-depth interviews with several of the stakeholders. The major refinement of the goals was the commitment to a completely model-based development with Simulink/TargetLink and the central management of all development-related artefacts and data in the data backbone eASEE. In addition, the interviews showed that the main interest at present was the analysis and assurance of the maintainability of the built models.

### Define Activities and Tasks

As we refined the general goals to the maintainability of the models, we consider *maintenance* the top level activity we are interested in. For that, we use a standard decomposition of maintenance activities from IEEE Standard 1219 [88]. Furthermore, we extended the activity tree to match the MAN development process by adding two activities (*Model Reading* and *Code Generation*) that are specific for the model-based development approach.

### Define Quality Goals

As we already narrowed the general goals to the analysis and assurance of maintainability, our main quality goal is to support the maintenance activity. The most important activity is code generation, because if we cannot generate meaningful source code, the model is useless. Then we are highly interested to be able to quickly read and understand the models. Finally, also the easy test of the models and the resulting code is of high importance.

### Identify Affected Technologies and Artefacts

From the general and quality goals, we derive that the main artefacts that are affected are the Simulink/TargetLink models of the automotive functions. In addition, as we are interested also in testing the system, the resulting source code could be important as well. Furthermore, as MAN uses the data backbone, the additional data in the backbone can influence our quality goals.

### Analyse Relevant Materials

The material we analysed for building the quality model consists of three types:

1. Existing guidelines for Simulink/Stateflow
2. Scientific studies about model-based development
3. Expert know-how of MAN's engineers

Specifically, our focus lies on the consolidation of four guidelines available for using Simulink and Stateflow in the development of embedded systems: the MathWorks documentation [144], the MAN-internal guideline, the guideline provided by dSpace [55], the developers of the TargetLink code-generator and the guidelines published by the MathWorks Automotive Advisory Board (MAAB) [145]. There is now also a MISRA guideline for Simulink [152] and TargetLink [153] which were not available when we built the model.

### Define Product Factors

Because of confidentiality reasons, we are not able to fully describe the MAN-specific model here. However, we present a number of examples that illustrate parts of the model. Overall, we modelled 87 product factors (64 new entities) that describe properties of entities not found in classical code-based development. Examples are states, signals, ports and entities that describe the graphical representation of models, e.g. colours.

We started with a simple translation of the existing MAN guidelines for Stateflow models into the maintainability model. For example, the MAN guideline requires the current state of a Stateflow chart to be available as a measurable output. This simplifies testing of the model and improves the debugging process. In terms of the model, this is expressed as [Stateflow Chart | ACCESSIBILITY] $\xrightarrow{+}$ [Debugging] and [Stateflow Chart | ACCESSIBILITY] $\xrightarrow{+}$ [Test].

We describe the ability to determine the current state with the property ACCESSIBILITY of the entity Stateflow Chart. The Stateflow chart contains all information about the actual statechart model. Note that we carefully distinguish between the *chart* and the *diagram* that describes the graphical representation. In the model the facts and impacts have additional fields that describe the relationship in more detail. This descriptions are included in generated guideline documents.

### Specify Quality Requirements

Finally, we classified most of the product properties in properties that *must not* or *should not* hold. This classification is sufficient for generating review guidelines and using the results for a manual maintainability analysis.

### 5.2.3 *Effects*

Consolidation of the Terminology

At MAN, we found that building a comprehensive quality model has the beneficial side effect of creating a consistent terminology. By consolidating the various sources of guidelines, we discovered a very inconsistent terminology that hampers a quick understanding of the guidelines. Moreover, we found that even at MAN the terminology has not been completely fixed. Fortunately, building a quality model automatically forces the modeller to give all entities explicit and consistent names. The entities of the facts tree of our maintainability model automatically define a consistent terminology and thereby provide a glossary.

One of many examples is the term *subsystem* that is used in the Simulink documentation to describe Simulink's central means of decomposition. The dSpace guideline, however, uses the same term to refer to a *TargetLink subsystem* that is similar to a Simulink subsystem but has a number of additional constraints and properties defined by the C-code generator. MAN engineers, on the other hand, usually refer to a *TargetLink subsystem* as *TargetLink function* or simply *function*. While building the maintainability model, this discrepancy was made explicit and could be resolved.

Resolution of Inconsistencies

Furthermore, we are able to identify inconsistencies not only in the terminology but also in contents. For the entity *Implicit Event*, we found completely contradictory statements in the MathWorks documentation and the dSpace guidelines.

- *MathWorks [144]* "Implicit event broadcasts [...] and implicit conditions [...] make the diagram easy to read and the generated code more efficient".
- *dSpace [55]* "The usage of implicit events is therefore intransparent concerning potential side effects of variable assignments or the entering/exiting of states".

Hence, MathWorks sees implicit events as improving the readability, while dSpace calls them intransparent. This is a clear inconsistency. After discussing with the MAN engineers, we adopted the dSpace view.

Revelation of Omissions

An important feature of the quality meta-model is that it supports inheritance. This became obvious in the case study after modelling the MAN guidelines for Simulink variables and Stateflow variables. We model them with the common parent entity Variable that has the attribute LOCALITY that expresses that variables must have the smallest possible scope. As this attribute is inherited by both types of variables, we found that this important property is not expressed in the original guideline. Moreover, we see by modelling that there was an imbalance between the Simulink and Stateflow variables. Most of the guidelines related only to Simulink variables. Hence, we transferred them to Stateflow as well.

#### Integration of Empirical Research Results

Finally, we give an example of how a scientific result can be incorporated into the model to make use of new empirical research. The use of Simulink and Stateflow has not been intensively investigated in terms of maintainability. However, especially the close relationship between Stateflow and the UML statecharts allows to reuse results. A study on hierarchical states in UML statecharts [44] showed that the use of hierarchies improves the efficiency of understanding the model in case the reader has a certain amount of experience. This is expressed in the model as follows: [Stateflow Diagram | STRUCTUREDNESS] $\xrightarrow{+}$ [Model reading].

### 5.2.4 *Usage of the Model*

At MAN, we concentrated on checklist generation and preliminary automatic analyses. We chose them, because they promised the highest immediate pay-off. The Quamoco tool chain did not yet exist.

#### Checklist Generation

We see quality models as central knowledge bases for quality issues in a project, company or domain. This knowledge can and must be used to guide development activities as well as reviews. The model in its totality, however, is too complex to be comprehended entirely. Hence, it cannot be used as a quick reference. Therefore, we exploit the tool support for the quality model to select subsets of the model and generate concise guidelines and checklists for specific purposes.

The MAN engineers perceived the automatic generation of guideline documents to be highly valuable as we could structure the documents to be read conveniently by novices as well as experts. Therefore, the documents feature a very compact checklist-style section with essential information only. This representation is favoured by experts who want to ensure that they comply to the guideline but do not need any further explanation. For novices the remainder of the document contains a hyperlinked section providing additional detail. Automatic generation enables us to conveniently change the structure of all generated documents. More importantly, it ensures consistency within the document which would be defect prone in handwritten documents.

Automatic Analyses

As the model is aimed at breaking down product factors to a level where they can be evaluated and they are annotated with the degree of possible automation, it is straightforward to implement automatic analyses. For the product properties that can be automatically evaluated, we were able to show that we can check them in Simulink and Stateflow models.

For this, we wrote a parser for the proprietary text format used by Matlab to store the models. Using this parser we are able to determine basic size and complexity metrics of model elements like states or blocks, for example. Moreover, we can use the parser to automatically identify model elements that are not satisfactorily supported by the C-code generator. We also implemented clone detection specifically for Simulink and Stateflow models and included it into the quality assessment from the model.

By integrating these analyses in our quality controlling toolkit ConQAT [50], we are able to create aggregated quality profiles and visualisations of quality data. We have not yet used the integrated quality evaluation approach described in Sect. 4.2. This would be the next logical step.

### 5.2.5 *Lessons Learned*

Overall, the MAN engineers found the approach of building the model for maintainability useful. Especially the model's explicit illustration of impacts on activities was seen as beneficial as it provides a sound justification for the quality rules expressed by the model. Moreover, the general method of modelling – that inherently includes structuring – improved the guidelines: Although the initial MAN guideline included many important aspects, we still were able to reveal several omissions and inconsistencies. Building the model, similar to other model building activities in software engineering [180], revealed these problems and allowed to solve them.

Another important result is that the maintainability model contains a consolidated terminology. By combining several available guidelines, we could incorporate the quality knowledge contained in them and form a single terminology. We found terms used consistently as well as inconsistent terminology. This terminology and combined knowledge base were conceived useful by the MAN engineers.

Although the theoretical idea of using an explicit quality meta-model for centrally defining quality requirements is interesting for MAN, the main interest is in the practical use of the model. For this, the generation of purpose-specific guidelines was convincing. We not only built a model to structure the quality knowledge, but we are able to communicate that knowledge in a concise way to developers, reviewers and testers. Finally, the improved efficiency gained by automating specific assessments was seen as important. The basis and justification for these checks are given by the model.

Table 5.4 Sample projects [138]

| Project | Description |
|---|---|
| A | Public sector, desktop application |
| B | Private partner, desktop application |
| C | Private partner, web application |
| D | Private partner, software component |
| E | Private partner, web application |
| F | Public sector, software component |

## 5.3 Capgemini TS

Based on the experiences with modelling maintainability at MAN Truck and Bus, we stepped onto new terrain with modelling security at Capgemini TS [138] who has its main focus on the individual development of business information systems.

### 5.3.1 *Context*

Capgemini TS is the technology service entity of Capgemini. Their main focus is on custom business information systems and therefore the systems are very different from what we encountered at MAN Truck and Bus.

As at Capgemini TS there are numerous projects, we cannot immediately build a quality model valid for all those projects. We therefore decided to restrict ourselves to a small initial sample for the model building. We interviewed contact persons from 20 projects. From this analysis, we found six projects (A–F) that were suitable for building the quality model since they were able to offer sufficient data. The sizes of the selected projects range from one person month to 333 person years; the average size was 163 person years (Table 5.4).

### 5.3.2 *Building the Model*

Define General Goals

Similarly to the huge number of projects and their diversity at Capgemini TS, there are various general goals by the existing stakeholders. In this particular case, we restricted ourselves to the business goal to ensure customer satisfaction by protecting their valuable assets. The business information systems of Capgemini TS often include large amounts of partly very sensitive information of their customers as well as perform vital services which are often exposed to the Internet. The protection of this data and services is essential to keep the customer satisfied.

### Analyse Documents Relevant for General Goals

The aim of this step is to refine the general goals to derive relevant activities and tasks. The protection of valuable assets is what we usually describe by *security*. Therefore, the main document we use as basis for the quality model is the security model from Sect. 2.6.2. It defines as activity hierarchy a classification of attacks on software systems which we can reuse for the Capgemini quality model. In addition, we compare the attacks from the security model with the existing requirements specifications of the analysed projects to remove irrelevant attacks or detect omissions in the activities. Capgemini TS especially uses the German *IT-Grundschutz* manual [29] of the BSI as a basis for security requirements.

### Define Activities and Tasks

The basic activity hierarchy is the same as in the model in Sect. 2.6.2. In particular, we extended it by the activities that we found in the *IT-Grundschutz* manual. It defines threats, such as the abuse of user rights, which we incorporated into the activity hierarchy.

### Define Quality Goals

We restricted ourselves in the Capgemini quality model early on to security-related activities. We have the quality goal that all attack activities in our activity hierarchy that are relevant for a particular software system are hard to perform. For example, the abuse of user rights needs to be difficult.

### Identify Affected Technologies and Artefacts

We then analysed what technologies and artefacts we need to describe by product properties to reach these goals. In the context of Capgemini TS, we needed to look at the programming and markup languages Java, JSP, ASP.NET, JavaScript, HTML and CSS. All artefacts created in these languages can be subject to attacks. Furthermore, the user and rights management is a key artefact in a software system for ensuring its robustness against attacks. Finally, although it might not be directly a part of the analysed software system, there is a network and operating system that influence attacks. Finally, there can be additional security systems such as firewalls.

### Analyse Relevant Material

To find relevant product factors for the technologies and artefacts, we go back to the material we have already analysed for activities. The security model from

Sect. 2.6.2 contains product factors and impacts to attacks from existing collections. The *IT-Grundschutz* manual describes *safeguards*, which are product properties to prevent threats: our attack activities. Finally, Capgemini TS has many experiences with building security-critical software systems. This expert knowledge from practical experience is also a valuable source for product factors. We, therefore, also analyse the existing requirements specifications of our sample projects.

### Define Product Factors

Corresponding to the threats, which we modelled as activities, the manual describes so-called *safeguards*, which are counter measures to these threats. We modelled the safeguards as product factors.

Overall, we defined several hundred product factors for the identified technologies and artefacts. It starts with general product factors that hamper attacks. For example, that passwords should never be saved or shown as clear text:
[Password | CONCEALMENT] $\xrightarrow{-}$ [Attack]

Then we described product factors for specific technologies. For example, in most Capgemini TS systems, there are database systems. We defined that instead of dynamically built SQL, a developer should use:

- An OR mapper
- Prepared statements
- Static SQL statements

[Database Access | APPROPRIATENESS] $\xrightarrow{-}$ [SQL injection]

Another example is the session handling in web applications. We defined that URLs do not contain session IDs and that each session has a time-out. Both have a negative impact on hijacking the sessions:
[URL | APPROPRIATENESS] $\xrightarrow{-}$ [Session Hi-Jacking]
[Session Length | LIMITEDNESS] $\xrightarrow{-}$ [Session Hi-Jacking]

As final example, we defined product factors that relate to the network connections of the software system. For valuable assets, such as sensitive user data, it works against many attacks to encrypt the transfer of these assets over networks:
[Asset Transfer | GUARDEDNESS] $\xrightarrow{-}$ [Attack]

### Specify Quality Requirements

As last step, some of the product factors need to be refined to concrete quality requirements. For example, [Session Length | LIMITEDNESS] $\xrightarrow{-}$ [Session Hi-Jacking] is not complete as it only prescribes that there needs to be some limit of the session length. In a quality requirement for a concrete system, we gave a session limit of 5 min. Other product factors, such as the concealment of passwords, could be directly used as quality requirements.

Table 5.5 Reuse potential [138]

| Project | # Sec. Reqs. | Reuse ratio |
|---|---|---|
| A | 127 | – |
| B | 23 | 0.87 |
| C | 48 | 0.27 |
| D | 5 | 0.60 |
| E | 29 | 0.52 |
| F | 408 | 0.18 |
| Mean | 106.67 | 0.47 |

### 5.3.3 *Effects*

Comprehensive and Concise Specification

We could improve the specification documents since the structure of the model prescribes a uniform way to document each requirement with two major effects:

1. Avoidance of redundancy
2. Explicit rationales

The clear structure prevents redundant or similar parts in general.

Reuse

Another important effect is that the quality model can act as a repository for security or other quality requirements. This central repository ensures requirements specifications with a high quality that can be reused and thereby reduce the possibility of defects in the requirements. We deliberately built the quality model from analysing one project after another. This way, we could analyse the overlap between the security requirements of the projects. This overlap is the potential reuse that can be realised by exploiting the quality model. Table 5.5 shows the results for our sample projects. The column *# Sec. Reqs* gives the number of security requirements in the quality model, and *Reuse Ratio* shows the share that could have been reused. The range of the size of the requirements is high which stems from the diversity of the analysed systems and their specifications. On average more than 100 security requirements were specified per project.

We found that on average 47 % of the security requirements could have been reused using the quality model. The range goes from 0.18 up to 0.87. The ratio, however, depends strongly on the size of the specification. In smaller specifications, it is by far easier to come to a high reuse ratio. With an average reuse ratio of 47 %, almost every second specified requirement could have been reused employing the quality model. Still, as we have no baseline of reuse that would be possible without the model approach, we cannot give an improvement caused by the approach but only show its potential. Among other factors, the size of the specification has a large effect on the reuse ratio. The model repository can only deliver as many requirements as are contained in it. Hence, the reuse potential for a large specification seems to be lower. Another effect, especially found in project F, is that larger specifications tend to have more redundancies. Because of the fixed structure of requirements in the model, this redundancies can largely be reduced.

Beyond these findings, we found further aspects to be discussed. First, in several cases we could reveal omissions in the specified security requirements. We derived additional requirements that were not contained in the original requirements specifications. Finally, we observed during the case study that the efforts of applying the model approach decreased while including the requirements of more and more projects into the repository. A major cause was easy access to quality requirements via goals. On the long run, such a central quality model can contribute to an efficient quality knowledge transfer to companies.

### 5.3.4 *Usage of the Model*

Capgemini TS uses the quality model very differently from MAN. While MAN focuses on the evaluation of their Simulink/TargetLink models using the quality model, Capgemini aimed at using the quality model to elicit and specify quality requirements. Both activities are closely related but can have positive effects independently.

### 5.3.5 *Lessons Learned*

Overall, also the engineers at Capgemini TS found the quality model approach useful to structure and reuse their quality requirements. Even though the quality evaluation possibilities are not exploited, the quality model is already useful as a means for structuring quality requirements. It ensures

1. the completeness of the quality requirements by employing existing experiences and
2. the existence of rationales for each requirements in the form of the activities, or attacks, it influences.

## 5.4 Telecommunications Company

A further practical experience, I want to report on, is about introducing a quality model and especially a dashboard at a company. This was part of an investigation of introducing quality control at a telecommunications company [67].

### 5.4.1 Context

The company we worked with in this case was a German telecommunications company which provides Internet, phone and TV services to end users. We had a small collaboration on improving and introducing quality management and control techniques in their software development process. Their software systems mainly consist of internal billing systems as well as web applications for the customers.

The idea of the collaboration was, among others, to understand the current state of software development in that company, find problems related to quality control, derive a suitable quality model and implement it using corresponding quality assurance techniques and measures. This collaboration took place before the Quamoco project and, hence, we were not able to use the fully fleshed-out Quamoco approach and tool chain. Instead, we concentrated on building an activity-based quality model to understand what data should be collected and implementing corresponding analyses. The results were supposed to be applicable in principle to all software systems of the company, but we concentrated on a specific larger system as an example.

### 5.4.2 Building the Model

As this collaboration had been before the Quamoco approach and the actual scope was a bit bigger, we did not follow the model building approach from Sect. 3.1 directly. Instead, we started with an interview to investigate the state of the software development at the company and learn about current quality problems. We interviewed 12 people from development, project management, requirements engineering, quality assurance and operation.

We found that the company successfully works with a mostly waterfall-like process with specification, design, implementation and testing sequentially. In this waterfall, there is an established procedure for quality gates after each phase. Only risk management is missing from the process description. In the testing phase, they start with unit tests which are done by the developers. Later integration and system tests are done by the quality assurance team. During development, they already used code reviews and coding guidelines as well as, in some projects, test-driven development. Especially the use of automated unit tests varied strongly over the different projects. If automated, tests are highly used for regression testing which is seen as very beneficial because problems are detected early. The test progress is controlled by the number of succeeded and failed test cases as well as by tracking assigned, resolved and closed change requests.

We discussed and proposed several measures and asked for the interviewees' opinions. The measures considered very important were test coverage, degree of dependency between modules, structuredness of the code, loading time of the program, code clones and execution time. Only of medium importance were

| | | | Activity | | | | | | | | | |
|---|---|---|---|---|---|---|---|---|---|---|---|---|
| | | | Maintenance | | | | | | Use | | Testing | |
| | | | Implementation | | | | Analysis | | | | | |
| Code | | | Interface Change | Refac. | Defect Remov. | Func. Impl. | Impact Analysis | Fault Local. | Learn | Normal Use | Regr. Testing | Testing new Func. |
| Static | Identifier | Consistency | + | + | + | + | + | + | | | | |
| Static | Comment | Suitability | + | + | + | + | + | + | | | | |
| Static | Unused Code | Existence | | | | | - | - | | | | |
| Static | Cyclom. Complex. | Extent | | - | - | | - | - | | | | |
| Static | Cloned Code | Existence | - | - | - | | - | - | | | | |
| Dynamic | Autom. Test Case | Existence | | | | | | + | | + | + | + |
| Dynamic | Autom. Test Case | Coverage | | | | | | + | | + | | |
| Dynamic | Cap./Rep. Test Case | Existence | | | | | | | | + | + | + |
| Dynamic | Manual Test Case | Existence | | | | | | | + | + | + | + |

**Fig. 5.4** The derived activity-based quality model for the telecommunication company

considered the comment ratio (code comments to code), dead code (because it is not important at run time) and number of classes or interfaces. Also size measures in general were not considered important at all apart from giving a general impression of the size of the system. Finally, we also let them classify quality factors to derive a corresponding quality model. They judged functional suitability, security, reliability, usability and modifiability as most important.

We used the information from the interviews to derive an activity-based quality model and corresponding measures to evaluate the product factors in the model. As shown in Fig. 5.4, we modelled if a product factor (left-hand side) has a positive (+) or negative (−) impact on the activities (top). We did not implement a complete evaluation aggregation along these pluses and minuses. Instead, we defined measures for the product factors and informally assumed that good measurement values for the product factors mean a good or bad impact on the activities.

As you can also see in Fig. 5.4, we concentrated on code as the root entity and structured it into dynamic aspects of the code and static aspects. In the static aspects, we included the entities Cloned Code, Cyclomatic Complexity, Unused Code, Comment and Identifier. For the dynamic aspects, we included Manual Test Case, Capture/Replay Test Case and Automated Test Case. In the Quamoco approach, we would probably model the entities differently, for example, using Method and COMPLEXITY as its property to form a product factor. This quality model shows, however, that it is not necessarily important to follow the modelling principles of Quamoco closely and still derive benefit from building an explicit quality model.

As the entities were already rather specific, most of their properties are only EXISTENCE, but we also used CONSISTENCY, SUITABILITY, EXTENT and COVERAGE. Following from the important quality characteristics, we included the activities Maintenance, Use and Testing in the quality model. Each of these activities were broken down in several sub-activities to make the impacts from product factors more clear. For example, the [Cyclomatic Complexity | EXTENT] has a negative influence on Refactoring and Defect Removal but not the Functionality Implementation of new features.

Finally, we had to define measures for each of the product factors together with thresholds for what is considered good or bad. In the Quamoco approach, we would map this to a linear increasing or decreasing function to fully operationalise the quality evaluation. Here, we just implemented the analysis in ConQAT and showed red, yellow or green traffic lights for each measure. For example, for the comment ratio, we considered a ratio between 0 % and 15 % as well as between 85 % and 100 % as red, between 15 % and 30 % as well as 70 % and 85 % as yellow and between 30 % and 70 % as green. Similarly, we defined thresholds for measures such as clone coverage, architecture conformance, cyclomatic complexity and test coverage.

### 5.4.3 Effects

After the introduction of the measures and analyses, we conducted a survey among the previously interviewed engineers and beyond to judge the benefit generated. They saw very high benefit in better architecture specifications, triggered by the architecture conformance analysis, and additional dynamic tests, triggered by the monitoring of tests. Similarly, the engineers judged the consequent definition, conformance and updating of architecture specifications, test-driven development and overall the new static analysis as highly beneficial. They realised high benefits from analysing code clones, unit test results and test coverage. Only medium benefit brought the analysis of unused code and cyclomatic complexity which seemed to not add so much more additional information. Overall, most of the engineers answered that they check the analyses in the ConQAT dashboard every day; only some check them only every week.

### 5.4.4 Lessons Learned

We found that the developers were overall happy about the additional information about the quality of their systems that we derived using the activity-based quality model. Especially architecture specification and analysis was underdeveloped and, therefore, a welcome addition. The explicit modelling of relationships in the activity-based quality model helped to make the quality goals transparent. Today, we would use the Quamoco evaluation approach and toolset to support a complete operationalisation of the model. Nevertheless, even this rather pragmatic model helped in structuring the quality evaluation.

## 5.5 Siemens COM

We worked with the network communication division of Siemens on a very specific quality model and quality evaluation. The aim was to analyse and predict failures in the first year in the field. The quality control loop concentrated on the last phases, especially system test and field test. It shows the use of testing as data source for quality evaluation. We reported these experiences initially in [207].

### 5.5.1 *Context*

Siemens Enterprise Communications is now a joint venture of the The Gores Group and Siemens AG, but it was still part of Siemens AG at the time of our collaboration. The company has a strong history in voice communication systems and expanded into other areas of communication and collaboration solutions. They have offices worldwide and employ about 14,000 people.

We worked with the quality assurance department which is responsible for system and field testing. The aim was to analyse and optimise system tests and field tests by predicting the number of field failures in the first year of a new product release.

### 5.5.2 *Quality Model*

As we had a very concrete and focused goal, we did not build a broad Quamoco quality model to capture all potentially interesting product factors, but we identified the essential activities and properties. The overall goal was to minimise the disruptions of the users of the Siemens systems by software failures. Hence, in an activity-based quality model, our quality goal was to maximise the effectiveness of the interaction of the users with the system. In a product quality view from ISO/IEC 25010, we wanted to maximise reliability.

Similar to the experiences at MAN and Capgemini, we could have built a quality model that describes all product factors that influence the effectiveness of the interaction of the user, use that to specify detailed quality requirements and analyse the artefacts of the system. In this case, however, we had a very focused prediction aim, the number of field failures in the first year. To come to valid predictions, we decided to concentrate on a single product factor: faultiness. Faults in the product cause the failures we are interested in to predict. Therefore, we needed to estimate the number of faults and their failure rate to predict reliability. This gives us the measures we need. For a concrete prediction of the effectiveness of interaction or reliability, we needed a stochastic model that captures the relationship between faults and failures.

### 5.5.3 *Stochastic Model*

The description of the software failure process using stochastic models is a well-established part of software reliability engineering. We call those models commonly software reliability growth models, because we assume that over time software reliability grows as faults are removed by fixes. All kinds of stochastic means have been tried for describing the software failure process and for predicting reliability. You can find more details on any aspects of these models in a variety of books [139, 158].

For Siemens, we looked at four stochastic models that had shown to be good predictors in other contexts as well as a new stochastic model, which we developed specifically for Siemens based on the kind of relationship between faults and failures we found in their data.

#### Musa Basic

The Musa basic execution time model is one of the most simple and therefore easy to understand reliability growth models, because it assumes that all faults are equally likely to occur and are independent of each other and there is a fixed number of faults. Hence, it abstracts from introducing new faults, for example, by bug fixes. The intensity with which failures occur is then proportional to the number of faults remaining in the program and the fault correction rate is proportional to the failure occurrence rate.

#### Musa–Okumoto

The Musa–Okomoto model, also called logarithmic Poisson execution time model, was first described in [160]. It also assumes that all faults are equally likely to occur and are independent of each other. Here, the number of faults, however, is not fixed. The expected number of faults is a logarithmic function of time in this model, and the failure intensity decreases exponentially with the expected failures experienced. Hence, the software will experience an infinite number of failures in infinite time.

NHPP

There are various models based on a non-homogeneous Poisson process [172]. They all are all Poisson process which means that they assume that failure occurs independently. We do not model faults but the number of failure up to a specific point in time. These models are called *non-homogeneous*, because the intensity of the failure occurrence is changing over time.

Littlewood–Verall Bayesian

This model was proposed for the first time in [136]. It also does not take faults explicitly into account. Its assumptions are that successive times between failures are independent random variables each having an exponential distribution and each failure is then following a gamma distribution.

Fischer–Wagner

We developed the last model specifically for the context of Siemens and their communication systems. They had observed a geometric sequence (or progression) between failure rates of faults. NASA and IBM had already documented similar observations [1, 164, 165]. We used this relationship to develop a fifth stochastic model that included an estimation of faults. There are more details on this model available in [206, 207].

### 5.5.4 *Time Component*

A critical part of predicting reliability is to analyse time correctly. We are interested in the occurrence of failures in a certain time interval. For software, however, calendar time is usually meaningless, because there is no wear and tear. The preferable way is to use execution time directly. This, however, is often not possible. Subsequently, a suitable substitute must be found. With respect to testing this could be the number of test cases or for the field use the number of clients. Figure 5.5 shows the relationships between different possible time types.

The first possibility is to use in-service time as a substitute. This requires knowledge of the number of users and the average usage time per user. Then the question arises how this relates to the test cases in system testing. A first approximation is the average duration of a test case.

In the context of the Siemens communication systems, the most meaningful time are incidents, each representing a usage of the system. In the end, however, we want a prediction of failures over 1 year. To convert these incidents into calendar time

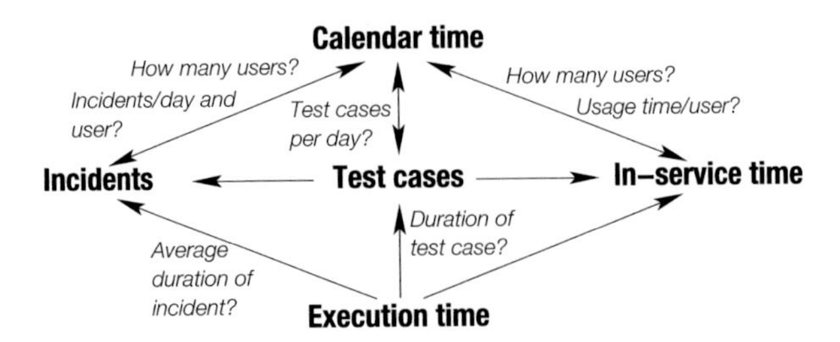


**Fig. 5.5** The possible relationships between different types of time [207]

it is necessary to introduce an explicit time component. It contains explicit means to convert from one time format into another.

The number of incidents is, opposed to the in-service time, a more task-oriented way to measure time. The main advantage of using incidents, apart from the fact that they are already in use at Siemens, is that in this way, we can obtain very intuitive metrics, e.g. the average number of failures per incident. There are usually some estimations of the number of incidents per client and data about the number of sold client licences.

### *5.5.5 Parameter Estimation*

The stochastic models describe the relationship of faults in the product to failures in the field. They rely, however, on parameters such as the number of faults to make a prediction of the reliability of the system. We need to estimate these parameters accurately to get a valid prediction.

There are two techniques for parameter determination currently in use. The first is prediction based on data from similar projects. This is useful for planing purposes before failure data is available. The second is for prediction during test, field trial, and operation based on the sample data available so far. This is the technique most reliability models use and it is also statistically most advisable since the sample data comes from the population we actually want to analyse. Techniques such as Maximum Likelihood estimation or Least Squares estimation are used to fit the model to the actual data.

For the application at Siemens we chose the Least Squares method for estimating the parameters of the models. In that method an estimate of the failure intensity is used and the relative error to the estimated failure intensity from the model is minimised. For our Fischer–Wagner model, we implemented it separately. The other models are implemented in the tool SMERFS [60] that we used to calculate the necessary predictions.

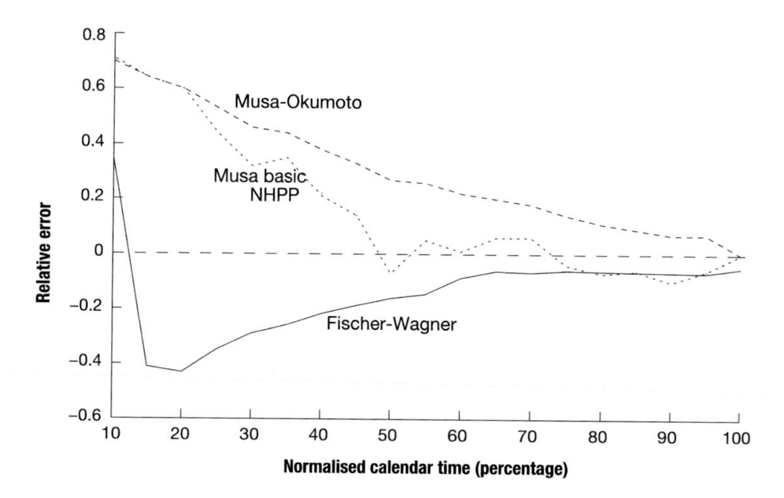


**Fig. 5.6** Relative error curves for the models based on the Siemens 1 data set [207]

## 5.5.6 *Suitability Analysis*

There is no one-size-fits-all reliability growth model. We always need to run different models to find the most suitable one for any given context. For this we relied on original data from Siemens from the field trial of two products. To anonymise the projects we call them *Siemens 1* and *Siemens 2*.

We followed [159] and used the *number of failures approach* to analyse the validity of the models for the available failure data. We assume that there have been $q$ failures observed at the end of test time (or field trial time) $t_q$. We use the failure data up to some point in time during testing $t_e (\leq t_q)$ to estimate the parameters of the mean number of failures $\mu(t)$. The substitution of the estimates of the parameters yields the estimate of the number of failures $\hat{\mu}(t_q)$. The estimate is compared with the actual number at $q$. This procedure is repeated with several $t_e$s.

For the comparison we plot the relative error $(\hat{\mu}(t_q) - q)/q$ against the normalised test time $t_e/t_q$ (see Figs. 5.6 and 5.7). The error will approach 0 as $t_e$ approaches $t_q$. If the relative error is positive, the model tends to overestimate, and vice versa. Numbers closer to 0 imply a more accurate prediction and hence a better model.

## 5.5.7 *Results*

We give for each analysed project a brief description and show a plot of the relative errors of the different models.

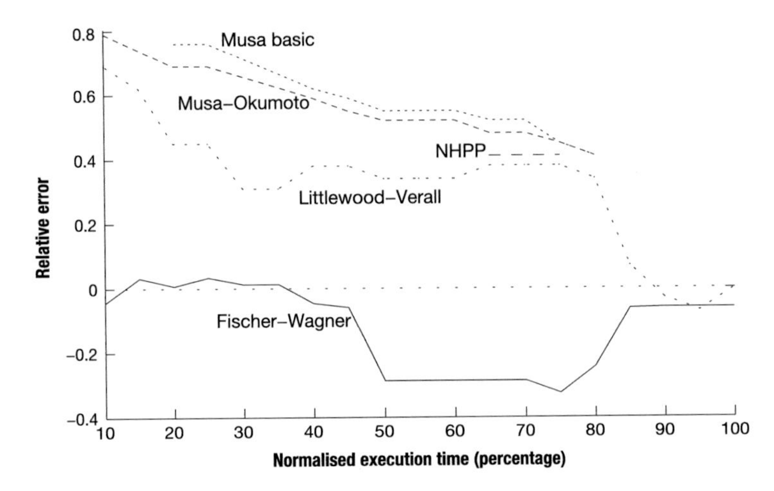


**Fig. 5.7** Relative error curves for the models based on the Siemens 2 data set [207]

#### 5.5.7.1 Siemens 1

This data comes from a large Siemens project that we call *Siemens 1* in the following. The software is used in a telecommunication equipment.

We only look at the field trial because this gives us a rather accurate approximation of the execution time which is the actually interesting measure regarding software. It is a good substitute because the usage is nearly constant during field trial. Based on the detailed dates of failure occurrence, we cumulated the data and converted it to time-between-failure (TBF) data. This was then used with the Fischer–Wagner, Musa-basic, Musa–Okumoto and NHPP models. The results can be seen in Fig. 5.6. In this case we omit the Littlewood–Verall which made absurd predictions of over a thousand future failures.

The Musa-basic and the NHPP models yield similar results all the time. They overestimate in the beginning and slightly underestimate in the end. The Musa–Okumoto model overestimates all the time; the Fischer–Wagner model underestimates. All four models make principally usable predictions close to the real value from about half the analysed calendar time. The Fischer–Wagner model has a relative error below 0.2 from 45 % on, the Musa basic and the NHPP models even from 40 % on.

#### 5.5.7.2 Siemens 2

*Siemens 2* is a web application for which we only have a small number of field failures. This makes predictions more complicated as the sample size is smaller.

It is interesting to analyse, however, how the different models are able to cope with this. For this, we have plotted the results in Fig. 5.7.

Again not all models were applicable to this data set. The NHPP model only made predictions for a small number of data points; the Musa basic and the Musa–Okumoto models were usable mainly in the middle of the execution time. All models made comparably bad predictions as we expected because of the small sample size. Surprisingly, the Fischer–Wagner model did well in the beginning but worsened in the middle until its prediction became accurate in the end again. Despite this bad performance in the middle of the execution time, it is still the model with the best predictive validity in this case. Only the Littlewood–Verall model comes close to these results. This might be an indication that the Fischer–Wagner model is well suited for smaller sample sizes.

### 5.5.8 Lessons Learned

The experience in quality control with Siemens was special as we did not aim to install a broad quality model in a quality control loop but to analyse a very focused area. The aim was to optimise and control system and field testing. For this, we installed only a very focused subset of an Quamoco quality model: product faults and failures which are a reduction of the interaction between the user and the system. Furthermore, we had the clear goal to have a concrete quantification of a prediction of these failures.

We employed the existing work on software reliability growth models to select and build stochastic models that can quantify the relationship between faults and failures. Based on data from system and field testing, we were able to predict the future distribution of failures. We compared several stochastic models to get the best predictive validity.

The predictive validity we observed for two Siemens systems was not extremely high with only the data from initial testing. Most models, however, showed that they can predict reasonably accurately with enough data. Especially the context-specific model performed well.

Overall, such a concrete quantification of quality goals needs investment in data collection and stochastic analysis. For reliability, there are existing models and tools which make their application easier. For other quality goals, this is probably more difficult. But even for reliability, a basic understanding of statistics and these models is necessary to successfully apply them. Especially, a suitable time component that allows an intuitive notion of time for the system under analysis is essential. If these problems are handled, however, this approach provides good predictions for planning and optimisation.

## 5.6 Five SMEs

Small- and medium-sized enterprises (SMEs) are important in the global software industry. They usually have not the same level of resources available for quality control. Hence, it is especially important to support them with (semi-)automatic techniques. We introduced and evaluated a set of static analysis tools in five SMEs. The following is based on our existing report [75] where you can also find more details.

### *5.6.1 Context*

Experiences from technology transfer are only useful with information about the corresponding contexts. Therefore, we illustrate the SMEs which took part in the technology transfer, the employed static analysis techniques and the software systems we used as example applications for the transfer.

Before the technology transfer project, we had been in discussions with various SMEs about software quality control in general and newer analyses techniques such as clone detection. We realised that there is a need as well as an opportunity for SMEs and quality control because of its high potential for automation. Therefore, after discussions and workshops with several SMEs, we started a collaboration with five SMEs from the Munich area. Our goal was to transfer the static analysis techniques *bug pattern detection*, *clone detection* and *architecture conformance* to the SMEs and document our experiences.

Following the definition of the European Commission [58], one of the participating SMEs is micro-sized, two are small and two are medium-sized considering their number of employees and annual turnover. The SMEs cover various business and technology domains including corporate and local government controlling, form letter processing as well as diagnosis and maintenance of embedded systems. Four of them are involved in commercial software development, one in software quality assurance and consulting only. Hence, the latter could not provide an own software project.

Following the suggestion of the partner without a software project, we instead chose the humanitarian open source system OpenMRS,[4] a development of the equally named multi-institution, non-profit collaborative. This allowed us to analyse the five software systems briefly described in Table 5.6 in the technology transfer project. For some information, we had to contact the core developers of Open MRS directly who were responsive to our requests thankfully.

The analysed software systems contain between 100 and 600 kLOC. The developments of the systems 1–4 were done by the SMEs themselves and had

[4]http://www.openmrs.org

**Table 5.6** Analysed systems [75]

| System | Platform | Sources | Size [kLOC] | Business domain |
|---|---|---|---|---|
| 1 | C#.NET | Closed, commercial | $\approx$100 | Corporate controlling |
| 2 | C#.NET | Closed, commercial | $\approx$200 | Embedded device maintenance |
| 3 | Java | Open, non-profit | $\approx$200 | Health information management |
| 4 | Java | Closed, commercial | $\approx$100 | local government controlling |
| 5 | Java | Closed, commercial | $\approx$560 | Document processing |

started at most 7 years earlier. The project teams contained less than ten persons. The development of the systems 1 and 2 had already been finished before our project started.

### 5.6.2 Introduction of Static Analysis

We used a carefully developed procedure to introduce the static analysis techniques and gather corresponding experiences in the SME context.

First, we conducted a series of workshops and interviews to convince industrial partners to participate in our project and to understand their context and their needs. In an early information event, we explained the general theme of transferring QA techniques and proposed first directions. With the companies that agreed to join the project, we conducted a kick-off meeting and a workshop to create a common understanding, discuss organisational issues and plan the complete schedule. In addition, the partners each presented a software system that we could analyse as well as their needs concerning software quality. To intensify our knowledge of these systems and problems, for each partner we performed an interview with two interviewers and a varying number of interviewees. We then compared all interview results to find commonalities and differences. Finally, we had two consolidation workshops to discuss our results and plan further steps.

Second, we retrieved the *source code* for at least three versions of the systems, in particular major releases chosen by the companies, for applying our static analysis techniques. For bug pattern detection and architecture conformance analyses, we retrieved or built executables packed with debug symbols for each of these configurations. For architecture conformance we also needed an appropriate architecture documentation. To accomplish all that, the SMEs had to provide project data as far as possible including source code, build environment and/or debug builds as well as documentation of source code, architecture and project management activities.

Third, we applied each technique on the systems by running the tools and inspecting the results, i.e. findings and statistics. To accomplish this step, the partners had to provide support for technical questions by a responsible contact or by personal attendance at the meetings.

Fourth, we conducted a short survey using a prepared questionnaire to better gather the experiences and impressions of our collaboration partners at the SMEs. These helped us to further judge our experience and to set them into context.

### Bug Pattern Detection

By now, bug pattern detection is a rather conventional static analysis technique, although it is still not as widely used as we would expect considering its low effort needed and its long history. Therefore, we chose proven and available bug pattern detection tools in our technology transfer project. For Java-based systems, we used FindBugs 1.3.9 and PMD[5] 4.2.5. For the C#.NET systems, we used Gendarme[6] 2.6.0 and FxCop 10.0. We determined the tool settings during preliminary analysis test runs and experimented with different rule sets. Categories and rules which we considered as not important – based on discussion with the partners as well as requirements non-critical to the systems' application domains – were ignored during rule selection.

We chose to use a very simple classification of the rules to simplify classifying the found defects. We only distinguish between rules for *faults* (potentially causing failures), *smells* (simple to very complex heuristics for latent defects) and *minor* (less critical issues with focus on coding style).

For additional and language independent metrics (lines of code without comments, code-comment ratio, number of classes, methods and statements, depth of inheritance and nested blocks, comment quality) as well as for preparing results and visualising, we applied ConQAT.

We analysed the finding reports from the tool runs. This step involved the filtering of findings as well as the inspection of source code analysing the severity and our confidence of the findings and determining how to correct the found problems. To get feedback and to confirm our conclusions from the findings, we discussed them with our partners during a workshop.

### Code Clone Detection

We used the clone detection feature of ConQAT 2.7 for all systems. In case of conventional clone detection, the configuration consists of two parameters: the minimal clone length and the source code path. In case of gapped clone detection (see Sect. 4.4.2), additional gap-specific parameters such as maximal allowed number of gaps per clone and maximal relative size of a gap are required. Based on our earlier experiences and initial experimentation, we set the minimal clone length to ten lines of code, the maximal allowed number of gaps per clone to 1 and the

[5] http://pmd.sourceforge.net

[6] http://www.mono-project.com/Gendarme

maximal relative size of a gap in our analysis to 30 %. After providing the needed parameters, we ran the analysis.

To inspect the analysis metrics and particular clones, we used ConQAT. It provided a list of clones, lists of instances of a clone, a view to compare files containing clone instances and a list of discrepancies for gapped clone analysis. We employed this data to recommend corrective actions. Also in a series of runs of clone detection over different versions of respective systems, we monitored how several parameters evolve over time.

#### Architecture Conformance Analysis

We also used ConQAT for this technique. For each system, we first configured the architecture conformance part of ConQAT with the path to the source code and corresponding executables of the system. Then, we created the architecture models based on the architectural information given by the enterprises. In our case, ConQAT is only able to analyse *static* call relationships between components out of the box. Next, we ran the ConQAT architecture conformance analysis which compares the relationships given in the architecture model with the actual relationships in the code. At last, we analysed the found architecture violations and discussed them with our partners.

We used an architecture model consisting of hierarchical components and allowed and disallowed call relationships between these components. The modelled components needed to be mapped to code parts (e.g. packages, namespaces or classes) as basis for the automated analysis. We excluded code parts from the analysis that did not belong directly to the system (e.g. external libraries). ConQAT is then able to analyse the conformance of the system to the architecture model. Every existing relationship that is not allowed by the architectural rules represents a defect. The tool visualises defects both on the level of components and on the level of classes which allows a detailed and a more coarse-grained analysis. To eliminate tolerated architecture violations and to validate the created architecture models, we discussed every found defect with our partners at the SMEs. This allowed us to group similar defects and to provide a general understanding.

### 5.6.3 Experiences with Bug Pattern Detection

We made the experience that bug patterns are a powerful technique to gather a vast variety of information about potentially defective code. We also found, however, that it is necessary to carefully configure it specific for a project to avoid false positives and get the most out of it.

First, we had difficulties in determining the impact of findings on quality factors of interest and their consequences for the project (e.g. corrective actions, avoidance or tolerance). The rule categories, severity and confidence information by the tools helped but also were confusing sometimes.

Second, some rules exhibited many false positives either because their technical way of detection is fuzzy or because a precise finding is considered irrelevant in a project-specific context. The latter case requires an in-depth understanding of each of the rules, the impacts of findings and, subsequently, a proper classification of rules as minor or irrelevant. We neither measured the rates of false positives nor investigated costs and benefits thereof as our focus lay on the identification of the most important findings only.

Third, despite our workshops and discussions with our collaboration partners at the SMEs, we only had a limited view on the systems, their contexts and evolution. This, together with the limited support in the tools for selecting and filtering the applied rules, hampered our efforts in configuring the analyses appropriately.

We addressed the first two issues by discussing with our partners and selecting and filtering rules as far as possible so that they do not show irrelevant findings for the systems under analysis. To compensate the third issue, we had to put in manual effort for selecting the right rules. As most finding reports were technically well accessible, we utilised ConQAT to gain statistical information for higher-level quality metrics.

We achieved the initial setup of a single bug pattern tool in less than an hour. This step required knowledge about the internal structure of the system such as its directory structure and third party code. We used ConQAT to flexibly run the tools in a specific setting and for further processing of the finding reports. Having good knowledge of ConQAT, we completed the analysis setup for a system (selection of rules, adjustment of bug pattern parameters and framework setup) in about half a day.

The runs took between a minute and an hour depending on code size, rule selection and other parameters. Hence, bug pattern detection is definitely suitable to be included into automated build tasks. Part of the rules are computationally complex and some tools frequently required more than a gigabyte of memory, however. The manual effort after the runs can be split into review and reconfiguration. The review of a report took us from a few minutes up to half an hour. An overview of the effort can be found in Table 5.7.

As a result of the review of the analysis findings, we were able to identify a variety of defects of all categories and severities. We will not go into details about then numbers of findings, but we summarise the most important findings grouped by programming language [75]:

**C#** Among the rules with highest numbers of findings, FxCop and Gendarme reported *empty exception handlers*, *visible constants* and *poorly structured code*. There was only one consensually critical kind of finding related to correctness in system 2: unacceptable loss of precision through *wrong cast during an integer division* used for accounting calculations.

**Java** Among the rules with the highest numbers of findings, FindBugs and PMD reported *unused local variables*, *missing validation of return values*, *wrong use*

**Table 5.7** Effort spent per system for applying each of the techniques [75]

| Phase | Work step | Clone detection | Bug pattern detection | Architecture conformance |
|---|---|---|---|---|
| Introduction (configuration) and calibration | Analysis tools | $\leq 0.5$ h | $\leq 1$ h | $\leq 0.5$ h |
| | Aggregation via ConQAT | n/a | $\leq 0.5\,rmd$ | $\leq 0.5$ h |
| | Recalibration, $x$-times | n/a | $\leq x \cdot 0.5$ h | n/a |
| Application (analysis) | Run analysis | $\leq 5$ min | 1 min $\leq \cdot \leq$ 1 h | $\leq 10$ s |
| | Inspection of results | $\leq 1$ h, more for gapped CD | 5min $\leq \cdot \leq$ 0.5 h | 5 min $\leq \cdot \leq$ 0.5 h |

*of serialisable* and extensive *cyclomatic complexity*, *class/method size*, *nested block depth* and *parameter list*. There have only been two consensually critical findings, both in system 5, related to correctness: foreseeable *access of a null pointer* and an *integer shift beyond 32 bits* in a basic date/time component.

Independent of the programming language and concerning security and stability, we frequently detected the pattern *constructor calls an overwritable method* in four of the five systems and found a number of defects related to *error-prone handling of pointers*. Concerning maintainability, the systems contained *missing or unspecific handling of exceptions*, manifold *violation of code complexity metrics* and various forms of *unused code*.

In addition, we learned from the survey at that end of the project that all of the partners considered our bug pattern findings to be relevant for their projects. The sample findings we presented during our final workshop were perceived as being non-critical for the success of the systems but would have been treated if they had been found by such tools during the development of these software systems. The low number of consensually critical findings correlated well with the fact that the technique was known to all partners, that most of them had good knowledge thereof and regularly used such tools in their projects, i.e. at least monthly, at milestone or release dates. Nevertheless, three of them could gain additional insights into this technique. Overall, all of the enterprises decided to use bug patterns as an important QA technique in their future projects.

### 5.6.4 Experiences with Code Clone Detection

Code clone detection turned out to be the most straightforward and least complicated of the three techniques. It has some technical limitations, however, that could hinder its application in certain software projects. A major issue was the analysis of projects containing both markup and procedural code like JSP or ASP.NET. Since ConQAT supports either a programming language or a markup language during a single analysis, it is required to aggregate the results for both languages. To avoid this complication and to concentrate on the code implementing the application logic, we took into consideration only the code written in the programming language and ignored the markup code. Nevertheless, it is still possible to combine the results of clone detection of the code written in both languages to get more precise results.

Another technical obstacle was filtering out generated code from the analysed code basis. In one system, large parts of the code were generated by the parser generator ANTLR. We excluded this code from our analysis using ConQAT's feature to ignore code files specified by regular expressions.

The effort required to introduce clone detection is small compared to bug pattern detection and architecture conformance analysis. Introducing clone detection is easy because configuring it is simple. In the simplest case, it only needs the path to the source code and the minimal length of a clone.

For all systems, it took less than an hour to configure the clone detection, to get the first results and to investigate the longest and the most frequent clones. Running the analysis process itself took less than 5 min. In case of gapped clone detection, it could take a considerable amount of time to analyse if a discrepancy is intended or if it is a defect. To speed up the process, ConQAT supports that the intended discrepancies can be fingerprinted and excluded from further analysis runs. An overview of the efforts is shown in Table 5.7.

The results of conventional clone detection can be interpreted as an indicator of bad design or of bad software maintainability, but they do not point at actual defects. Nevertheless, these results give first hints which code parts must be improved. The following three design flaws were detected in all analysed systems to a certain extent: cloning of exception handling code, cloning of logging code and cloning of interface implementation by different classes.

Table 5.8 shows the clone detection results for three versions of each study object (SO), sorted by version. In the analysed systems, clone coverage (Sect. 4.4.2) varied between 14 % and 79 %. Koschke [128] reports on several case studies with clone coverage values between 7 % and 23 %. He also mentions one case study with a value of 59 %, which he defines as extreme. Therefore, the SOs 1, 3 and 5 contain normal clone rates according to Koschke. The clone rate in SO 2 is higher than the rates reported by Koschke, and for SO 4 it is extreme. Regarding maintenance the calculated blow-up for each system is an interesting value. It shows by how much the system is larger than it needs to be if the cloning would be removed. For example, version III of SO 4 is more than three times bigger as its equivalent system containing no clones. SO 4 shows that cloning can be an increasing factor over time, while SO 3 reveals that it is possible to reduce the amount of clones existing in the system code.

Interesting values are also the longest clones and the clones with the most instances. The longest clones show if only smaller code chunks are copied or whole parts of the system. Those largest clones are also interesting candidates for refactorings because they allow to reduce the most redundancy with only tackling a small number of clones. They are usually measured in *units* which are essentially

**Table 5.8** Results of code clone detection [75]

| SO | Version | Blow-up [%] | Clone coverage [%] | Longest clone [units] | Most clone instances |
|---|---|---|---|---|---|
| 1 | I | 119.5 | 22.2 | 112 | 39 |
| | II | 118.9 | 23.0 | 117 | 39 |
| | III | 119.2 | 24.0 | 117 | 39 |
| 2 | I | 143.1 | 40.5 | 63 | 64 |
| | II | 150.2 | 45.4 | 132 | 47 |
| | III | 137.4 | 36.7 | 89 | 44 |
| 3 | I | 114.5 | 18.2 | 79 | 21 |
| | II | 111.2 | 15.1 | 52 | 20 |
| | III | 110.0 | 13.7 | 52 | 19 |
| 4 | I | 238.8 | 68.0 | 217 | 22 |
| | II | 309.6 | 77.6 | 438 | 61 |
| | III | 336.0 | 79.4 | 957 | 183 |
| 5 | I | 122.3 | 24.8 | 141 | 72 |
| | II | 122.7 | 25.3 | 158 | 72 |
| | III | 122.8 | 25.5 | 156 | 72 |

statements in the code. Also interesting to investigate are the clones with the most instances. That means which piece of code has the most copies. Those clones are often the most dangerous because it is very easy to forget to make changes to all copies. Therefore, they are also interesting candidates for the first refactorings.

Cloning is especially considered harmful, because it increases the chance of unintentional, inconsistent changes which can lead to faults in a system [113]. These changes can be detected when applying gapped clone detection. We found a number of such changes in the cloned code fragments, but we could not finally classify them as defects, because we lacked the knowledge needed about the software systems. Also the project partners could not easily classify these discrepancies as defects which confirms that gapped clone detection is a more resource demanding type of analysis. Nevertheless, in some clone instances, we identified additional instructions or deviating conditional statements compared to other instances of the same clone class. Gapped clone detection does not go beyond method boundaries, since experiments showed that inconsistent clones that cross method boundaries in many cases did not capture semantically meaningful concepts [113]. This explains why metrics such as clone coverage may differ from values observed with conventional clone detection. Table 5.9 shows our results of gapped clone detection.

Following the feedback obtained from the questionnaire, two enterprises had limited prior experience with clone detection; the others did not know about it at all. Three enterprises estimated the relevance of clone detection to their projects as very high; the others estimated it as medium relevant. One company stated that “clones are necessary within short periods of development”. Finally, all enterprises evaluated the importance of using clone detection in their projects as medium to high and plan to introduce this technique in the future.

**Table 5.9** Results of gapped code clone detection [75]

| SO | Version | Blow-up [%] | Clone coverage [%] | Longest clone [units] | Most clone instances |
|---|---|---|---|---|---|
| 1 | I | 119.9 | 22.3 | 34 | 39 |
| | II | 117.9 | 21.5 | 37 | 52 |
| | III | 117.4 | 22.1 | 52 | 52 |
| 2 | I | 116.3 | 19.0 | 156 | 37 |
| | II | 123.2 | 25.0 | 156 | 37 |
| | III | 123.7 | 25.3 | 156 | 37 |
| 3 | I | 124.4 | 18.2 | 73 | 123 |
| | II | 120.0 | 15.1 | 55 | 67 |
| | III | 118.6 | 20.5 | 55 | 64 |
| 4 | I | 192.1 | 58.6 | 42 | 34 |
| | II | 206.2 | 59.8 | 51 | 70 |
| | III | 211.1 | 59.5 | 80 | 183 |
| 5 | I | 117.4 | 20.7 | 66 | 68 |
| | II | 118.0 | 21.3 | 85 | 78 |
| | III | 118.2 | 21.5 | 85 | 70 |

### 5.6.5 *Experiences with Architecture Conformance Analysis*

We observe two kinds of general problems that prevent or complicate each architectural analysis: the absence of an architecture documentation and the usage of dynamic patterns.

For two of the systems there was no documented architecture available. In one case the information was missing, because the project was taken over from a different organisation that was not documenting the architecture at all. They reasoned that any later documentation of the system architecture would be too expensive. In another case, the organisation was aware that their system was severely lacking any architectural documentation. Nevertheless, they feared that the time involved and the sheer volume of code to be covered exceeds the benefits. The organisation additionally argued that they are afraid of having to update the documentation within several months as soon as the next release is coming out.

In system 2, a dynamic architectural pattern is applied, where nearly no static dependencies could be found between defined components. All components belonging to the system are connected at run time. Thus, our static analysis approach could not be applied.

Architecture conformance analysis needs two ingredients apart from the architecture documentation: the source code and the executables of a system. This could be a problem because the source has to be compilable to analyse it. Another technical problem occurred when using ConQAT. Dependencies to components solely existing as executables were not recognised by the tool. For that reason all rules belonging to compiled components could not be analysed. Beside these

**Table 5.10** Architectural characteristics of the study objects [75]

| SO | Architecture | Version | Violating component relationships | Violating class relationships |
|---|---|---|---|---|
| 1 | 12 Components | I | 1 | 5 |
| | 20 Rules | II | 3 | 9 |
| | | III | 2 | 8 |
| 2 | Dynamic | n/a | n/a | n/a |
| 3 | Undocumented | n/a | n/a | n/a |
| 4 | 14 Components | I | 0 | 0 |
| | 9 Rules | II | 1 | 1 |
| | | III | 2 | 4 |
| 5 | Undocumented | n/a | n/a | n/a |

problems, we could apply our static analysis approach to two systems without any technical problems. An overview of all systems with respect to their architectural properties can be found in Table 5.10.

For each system, the initial configuration of ConQAT and the creation of the architecture model in ConQAT could be done in less than 1 h. Table 5.10 shows the number of modelled components and the rules that were needed to describe their allowed connections. The analysis process itself finished in less than 10 s. The time needed for the interpretation of the analysis results is dependent on the amount of defects found. For each defect, we were able to find the causal code parts within 1 min. We expect that the effort needed for bigger systems will only increase linearly but stay small in comparison to the benefit that can be achieved using architecture conformance analysis. An overview of the efforts can be found in Table 5.7.

As shown in Table 5.10, we observed several discrepancies in the analysed systems over nearly all version. Only one version did not contain architectural violations. Overall, we found three types of defects in the analysed systems. Each defect represents a code location showing a discrepancy to the documented architecture. The two analysable systems had architectural defects which could be avoided if this technique had been applied. In the following we explain the types of defects we classified together with the responsible enterprises. The companies rated all findings as critical.

- *Circumvention of abstraction layers:* Abstraction layers (e.g. a presentation layer) provide a common way to structure a system into logical parts. The defined layers are hierarchically dependent on each other, reducing the complexity in each layer and allowing to benefit from structural properties like exchangeability or flexible deployment of each layer. These benefits vanish when the layer concept is harmed by dependencies between layers that are not connected to each other. In our case the usage of the data layer from the presentation layer was a typical defect we found in the analysed systems.

- *Circular dependencies:* We found undocumented circular dependencies between two components. We consider these dependencies – whether documented or not – as defects themselves, because they affect the general principle of component design. Two components that are dependent on each other can only be used together and can thus be considered as one component, which contradicts the goal of a well-designed architecture. The reuse of these components is strongly restricted. They are harder to understand and to maintain.
- *Undocumented use of common functionality:* Every system has a set of common functionality (e.g. date manipulation) which is often grouped into components and used across the whole system. Consequently, it is important to know where this functionality is actually used inside a system. Our observation showed that there were such dependencies that were not covered by the architecture.

As a result from the final survey, we observed that four of the five participating enterprises did not know about the possibility of automated architecture conformance analysis. Only one of them already checked the architecture of their systems, however in a manual way and infrequently. Confronted with the results of the analysis, all enterprises rated the relevance of the presented technique relevant. One of them stated that as a new project member, it is easier to become acquainted with a software system if its architecture conforms to its documented specification. All enterprises agreed on the usefulness of this technique and plan its future application in their projects.

### 5.6.6 Lessons Learned

First, we observed that code clone detection and architecture conformance analysis have been quite new to our partners as opposed to bug pattern detection which was well known. This may result from the fact that style checking and simple bug pattern detection are standard features of modern development environments. We consider it as important, however, to know that code clone detection can indicate critical and complex relationships residing in the code at minimum effort. We also made our partners aware of the usefulness of architecture conformance analysis, both in the case of an available architecture specification and to reconstruct such a documentation.

Second, we conclude that all of the three techniques can be introduced and applied with resources affordable for small enterprises. We assume that, except for calibration phases at project initiation or after substantial product changes, the effort of readjusting the settings for the techniques stays very low. This effort is compensated by the time earned through narrowing results to successively more relevant findings. Moreover, our partners perceived all of the discussed techniques as useful for their future projects.

Third, we came across interesting findings from the analysed systems. We found large clone classes, a significant number of pattern-based bugs aside from smells and pedantry as well as unacceptable architecture violations.

In summary, in our opinion static analysis tools can efficiently improve quality assurance in SMEs, if they are continuously used throughout the development process and are technically well integrated into the tool landscape.

# Chapter 6
# Summary

Software quality is a very complex issue. It contains many different aspects we need to take care of in the development of current systems. There is no easy way to handle it. As always, there is no silver bullet. What we have, however, are proven best practices and innovative techniques to build quality in and fight quality decay.

We have proposed in this book that you build processes and infrastructure for comprehensive quality control with diverse techniques based on operationalised models. This way, you take control of the results of your work instead of just reacting to problems that hit you. I hope I have equipped you with the necessary weapons to fight back and prevent these problems.

In this last chapter, we use the quality control loop introduced in Sect. 4.1 to organise and summarise the topics I have explained. Finally, we close with a summary of further readings that either help you to delve into specific topics or set some of our topics into a broader context.

## 6.1 Continuous Quality Control

A key metaphor for controlling quality we have used in this book is the quality control loop. We discussed it in detail in Sect. 4.1 and you can find it depicted in Fig. 6.1. At its beginning, there is a quality model that describes quality for a software system and which requirements and quality engineers use to specify its quality goals.

The quality engineer uses these goals and derived requirements to describe how the software should look like to the developers. The developers build the software system with a specific level of quality. The quality engineer then employs quality analysis techniques, such as reviews or analysis tools, to determine this quality level. This is the feedback for the quality engineers which they compare with the set quality goals. If there are deviations, the quality engineer instructs the developer accordingly. The developer changes the software, which the quality engineer then analyses and compares to the goals, and a new cycle begins.

S. Wagner, *Software Product Quality Control*, DOI 10.1007/978-3-642-38571-1_6,

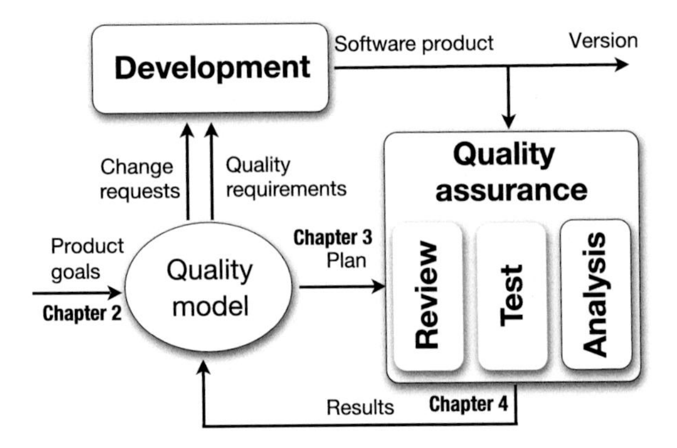


**Fig. 6.1** The book chapters in relation to the quality control loop

Chapter 2 lays the foundation for the control loop and concentrates on the quality model. As there are many definitions and types of quality models, we first set them into context and then described the international standard ISO/IEC 25010 and the Quamoco quality models in detail. In addition, similar to a software product, a quality model is not static but needs to change over time. We also need to make sure that our quality model is always up to date. We gave a small process that you should perform regularly. We illustrated the range of quality models with three explicit examples: the maintainability part of the Quamoco base model, a web security model and a reliability growth model.

Chapter 3 describes the planning part of quality control: model building and V&V planning. We have to build an appropriate quality model, usually by changing and extending an existing one, to be able to describe quality goals and specify concrete quality requirements. We also need to plan what quality assurance techniques we apply when. At best, we can derive this information from the quality model again.

Chapter 4 describes the main process: quality control. Therefore, it covers most of the control loop. The main topics, however, are in quality assurance techniques which are part of the quality analysis. We covered efficient general techniques such as reviews and inspections, testing and automatic static analysis as well as more specific techniques such as redundancy analysis and architecture conformance analysis. To aggregate and communicate the quality analysis results, we discussed the so-called dashboards which are a help in comparing the analysis results with the goals and keeping an overview of the various results. Finally, we showed how to improve quality overall, the quality of the control loop. For this, we briefly went into process improvement, which is also a basis for improving product quality.

Chapter 5, finally, covered concrete experiences with the techniques and processes we have discussed so far. We wanted, first, to demonstrate that they are not purely academic but also are applicable in practice and, second, help you in implementing them in your company by giving you concrete instances in real companies. These experiences cover different parts of the control loop in different depths. We have the broad Quamoco base model, two quality models we built using the model building process at MAN Truck and Bus and Capgemini TS and another activity-based quality model we built for a telecommunications company. We built a stochastic quality model for reliability analysis at Siemens. Finally, we introduced static analysis to five SMEs to help them implement the continuous control loop.

## 6.2 Further Readings

We have mentioned several books and papers worth reading for specific topics throughout the book. In the following, we summarise this for you to give you a starting point for your next readings.

### *6.2.1 Software Engineering*

- Sommerville [192]
  Sommerville's book is a classic on software engineering. It is a comprehensive introduction to topics partly covered in this book, such as testing, but describes the whole development process.
- Pfleeger and Atlee [171]
  Similar to Sommerville, this book gives a broad introduction to software engineering. It structures it rather differently, however, and makes thus a good additional source.

### *6.2.2 Quality Assurance and Quality Control*

- Garvin [71]
  Garvin does not talk specifically about software quality but product quality in general. Nevertheless, it gives a very interesting and plausible description of different views on software quality that reduces the confusion about quality.
- Parnas [170]
  In this paper, Parnas introduced "software ageing".
- Deissenboeck et al. [48]
  Our discussion on continuous quality control with a special focus on what tools can be used to support it.
- Grady and Caswell [77]
  A classic book on software measurement. It is still useful to help in establishing measurements.

### 6.2.3 Quality Models

- Kitchenham et al. [121]
  The SQUID approach to quality modelling was one of the first to support flexible modelling of quality.
- Wagner et al. [211]
  A compact description of the Quamoco approach to quality modelling and assessment.
- Musa [158]
  A very readable introduction to stochastic reliability modelling.

### 6.2.4 Defect Detection Techniques

- Runeson et al. [184]
  This is a well-written article about what empirical research so far has shown about different defect detection techniques.
- Ayewah et al. [5]
  A good overview article about the static analysis tool FindBugs for Java.
- Myers [161]
  The classic book on software testing still contains all the necessary basics.

# References


1. Adams, E.N.: Optimizing preventive service of software products. IBM J. Res. Dev. **28**(1), 2–14 (1984)
2. Al-Kilidar, H., Cox, K., Kitchenham, B.: The use and usefulness of the ISO/IEC 9126 quality standard. In: Proceedings of the International Symposium on Empirical Software Engineering (ISESE'05). IEEE Computer Society, Silver Spring (2005)
3. Alexander, I.: Misuse cases: Use cases with hostile intent. IEEE Softw. **20**(1), 58–66 (2003)
4. Avižienis, A., Laprie, J.C., Randell, B., Landwehr, C.: Basic concepts and taxonomy of dependable and secure computing. IEEE Trans. Dependable Secure Comput. **1**(1), 11–33 (2004)
5. Ayewah, N., Hovemeyer, D., Morgenthaler, J.D., Penix, J., Pugh, W.: Using static analysis to find bugs. IEEE Softw. **25**(5), 22–29 (2008)
6. Bakota, T., Hegedũs, P., Körtvélyesi, P., Ferenc, R., Gyimóthy, T.: A probabilistic software quality model. In: Proceedings of the 27th IEEE International Conference on Software Maintenance (ICSM'11). IEEE Computer Society, Silver Spring (2011)
7. Bansiya, J., Davis, C.G.: A hierarchical model for object-oriented design quality assessment. IEEE Trans. Softw. Eng. **28**(1), 4–17 (2002)
8. Basili, V., Donzelli, P., Asgari, S.: A unified model of dependability: Capturing dependability in context. IEEE Softw. **21**(6), 19–25 (2004)
9. Basili, V., Green, S., Laitenberger, O., Lanubile, F., Shull, F., Sørumgård, S., Zelkowitz, M.: The empirical investigation of perspective-based reading. Empir. Softw. Eng. **1**(2), 133–164 (1996)
10. Basili, V., Rombach, H.: The TAME project: Towards improvement-oriented software environments. IEEE Trans. Softw. Eng. **14**(6), 758–773 (1998)
11. Basili, V., Selby, R.: Comparing the effectiveness of software testing strategies. IEEE Trans. Softw. Eng. **SE-13**(12), 1278–1296 (1987)
12. Basili, V.R., Caldiera, G., Rombach, H.D.: Goal question metric paradigm. In: Marciniak, J.C. (ed.) Encyclopedia of Software Engineering, vol. 1. Wiley, New York (1994)
13. Beck, K.: Extreme Programming Explained. Addison Wesley, Reading (2000)
14. Beck, K.: Test Driven Development. By Example. Addison-Wesley Longman, Boston (2002)
15. Beizer, B.: Software Testing Techniques, 2nd edn. Thomson Learning, London (1990)
16. Belady, L.A., Lehman, M.M.: A model of large program development. IBM Syst. J. **15**(3), 225–252 (1976)
17. Beliakov, G., Calvo, T., Mesiar, R.: Guest editorial. Foreword to the special issue on aggregation operators. IEEE Trans. Fuzzy Syst. **15**(6), 1030–1031 (2007)
18. Beliakov, G., Pradera, A., Calvo, T.: Aggregation Functions: A Guide for Practicioners. Studies in Fuzziness and Soft Computing. Springer, Berlin (2007)



S. Wagner, *Software Product Quality Control*, DOI 10.1007/978-3-642-38571-1,

19. Bessey, A., Block, K., Chelf, B., Chou, A., Fulton, B., Hallem, S., Henri-Gros, C., Kamsky, A., McPeak, S., Engler, D.: A few billion lines of code later: Using static analysis to find bugs in the real world. Commun. ACM **53**(2), 66–75 (2010)
20. Blin, M.J., Tsoukiàs, A.: Multi-criteria methodology contribution to the software quality evaluation. Softw. Qual. J. **9**, 113–132 (2001)
21. Boegh, J., Depanfilis, S., Kitchenham, B., Pasquini, A.: A method for software quality planning, control, and evaluation. IEEE Softw. **16**(2), 69–77 (1999)
22. Boehm, B., Huang, L., Jain, A., Madachy, R.: The ROI of software dependability: The iDAVE model. IEEE Softw. **21**(3), 54–61 (2004)
23. Boehm, B.W.: Software Engineering Economics. Prentice Hall, Englewood Cliffs (1981)
24. Boehm, B.W., Brown, J.R., Kaspar, H., Lipow, M., Macleod, G.J., Merrit, M.J.: Characteristics of Software Quality. North-Holland, Amsterdam (1978)
25. Boehm, B.W., Brown, J.R., Kaspar, H., Lipow, M., MacLeod, G.J., Merrit, M.J.: Characteristics of Software Quality. TRW Series of Software Technology, vol. 1. North-Holland, Amsterdam (1978)
26. Broy, M.: Requirements engineering as a key to holistic software quality. In: Proceedings of the 21th International Symposium on Computer and Information Sciences (ISCIS 2006). Lecture Notes in Computer Science, vol. 4236, pp. 24–34. Springer, New York (2006)
27. Broy, M., Deissenboeck, F., Pizka, M.: Demystifying maintainability. In: Proceedings of the 4th Workshop on Software Quality (4-WoSQ), pp. 21–26. ACM Press, New York (2006)
28. Broy, M., Jonsson, B., Katoen, J.P., Leucker, M., Pretschner, A. (eds.): Model-Based Testing of Reactive Systems. Lecture Notes in Computer Science, vol. 3472. Springer, New York (2005)
29. BSI: BSI-Standard 100: IT-Grundschutz (2008)
30. Buhr, K., Heumesser, N., Houdek, F., Omasreiter, H., Rothermel, F., Tavakoli, R., Zink, T.: DaimlerChrysler demonstrator: System specification instrument cluster. http://www.empress-itea.org/deliverables/D5.1_Appendix_B_v1.0_Public_Version.pdf (2003). Accessed 15 Jan 2008
31. Calvo, T., Mayor, G., Mesiar, R. (eds.): Aggregation Operators. New Trends and Applications. Studies in Fuzziness and Soft Computing. Physica, Wurzburg (2002)
32. Canfora, G., Cimitile, A., Garcia, F., Piattini, M., Visaggio, C.A.: Evaluating advantages of test driven development: A controlled experiment with professionals. In: Proceedings of the ACM/IEEE International Symposium on Empirical Software Engineering (ISESE'06), pp. 364–371. ACM Press, New York (2006)
33. Chess, B., West, J.: Secure Programming with Static Analysis. Addison-Wesley, Reading (2007)
34. Chidamber, S.R., Kemerer, C.F.: A metrics suite for object oriented design. IEEE Trans. Softw. Eng. **20**(6), 476–493 (1994)
35. Chillarege, R.: Orthogonal defect classification. In: Lyu, M.R. (ed.) Handbook of Software Reliability Engineering, Chap. 9. IEEE Computer Society Press/McGraw-Hill, Silver Spring/New York (1996)
36. Chillarege, R., Bhandari, I.S., Chaar, J.K., Halliday, M.J., Moebus, D.S., Ray, B.K., Wong, M.Y.: Orthogonal defect classification – a concept for in-process measurements. IEEE Trans. Softw. Eng. **18**(11) (1992)
37. Chulani-Devnani, S.: Bayesian analysis of software cost and quality models. Ph.D. thesis, University of Southern California (1997)
38. Chung, L., Nixon, B.A., Yu, E., Mylopoulos, J.: Non-Functional Requirements in Software Engineering. Kluwer, Dordecht (1999)
39. Cielkowski, M., Laitenberger, O., Biffl, S.: Software reviews: The state of the practice. IEEE Softw. **20**(6), 46–51 (2003)
40. CMMI Product Team: CMMI for development, version 1.3. Technical Report CMU/SEI-2010-TR-033, Software Engineering Institute (2010)
41. Coleman, D., Lowther, B., Oman, P.: The application of software maintainability models in industrial software systems. J. Syst. Softw. **29**(1), 3–16 (1995)

42. Collofello, J.S.: Introduction to software verification and validation. SEI Curriculum Module SEI-CM-13-1.1. http://www.sei.cmu.edu/reports/89cm013.pdf (1988)
43. Common criteria for information technology security evaluation, version 3.1. Available Online at http://www.commoncriteriaportal.org/
44. Cruz-Lemus, J.A., Genero, M., Manso, M.E., Piattini, M.: Evaluating the effect of composite states on the understandability of UML statechart diagrams. In: Proceedings of the 8th International Conference on Model Driven Engineering Languages and Systems (MoDELS'05). Springer, Berlin (2005)
45. Davis, A.M.: Software Requirements: Objects, Functions, and States, 2nd edn. Prentice Hall, Englewood Cliffs (1993)
46. Deissenboeck, F.: Continuous quality control of long-lived software systems. Ph.D. thesis, Technische Universität München (2009)
47. Deissenboeck, F., Hummel, B., Juergens, E., Schaetz, B., Wagner, S., Girard, J.F., Teuchert, S.: Clone detection in automotive model-based development. In: Proceedings of the 30th International Conference on Software Engineering (ICSE'08), pp. 603–612. IEEE Computer Society, Silver Spring (2008)
48. Deissenboeck, F., Juergens, E., Hummel, B., Wagner, S., y Parareda, B.M., Pizka, M.: Tool support for continuous quality control. IEEE Softw. **25**(5), 60–67 (2008)
49. Deissenboeck, F., Juergens, E., Lochmann, K., Wagner, S.: Software quality models: Purposes, usage scenarios and requirements. In: Proceedings of the 7th International Workshop on Software Quality (WoSQ '09). IEEE Computer Society, Silver Spring (2009)
50. Deissenboeck, F., Pizka, M., Seifert, T.: Tool support for continuous quality assessment. In: Proceedings of the IEEE International Workshop on Software Technology and Engineering Practice (STEP), pp. 127–136. IEEE Computer Society, Silver Spring (2005). doi:http://doi.ieeecomputersociety.org/10.1109/STEP.2005.31
51. Deissenboeck, F., Wagner, S., Pizka, M., Teuchert, S., Girard, J.F.: An activity-based quality model for maintainability. In: Proceedings of the 23rd International Conference on Software Maintenance (ICSM '07). IEEE Computer Society, Silver Spring (2007)
52. Deming, W.E.: Out of the Crisis. MIT Press, Cambridge (2000)
53. Detyniecki, M.: Fundamentals on aggregation operators. In: Proceedings of the AOGP 2001 (2001) http://www-poleia.lip6.fr/~marcin/papers/Detynieck_AGOP_01.pdf
54. Dromey, R.G.: A model for software product quality. IEEE Trans. Softw. Eng. **21**(2) (1995)
55. dSpace: Modeling Guidelines for MATLAB/ Simulink/ Stateflow and TargetLink (2006)
56. Eick, S.G., Graves, T.L., Karr, A.F., Marron, J.S., Mockus, A.: Does code decay? Assessing the evidence from change management data. IEEE Trans. Softw. Eng. **27**(1), 1–12 (2001)
57. Endres, A., Rombach, D.: A Handbook of Software and Systems Engineering: Empirical Observations, Laws and Theories. The Fraunhofer IESE Series on Software Engineering. Pearson Education Limited, Harlow (2003)
58. European Commission: Commission recommendation of 6 May 2003 concerning the definition of micro, small and medium-sized enterprises. Off. J. Eur. Union **L 124**, 36–41 (2003)
59. Fagan, M.E.: Design and code inspections to reduce errors in program development. IBM Syst. J. **15**(3), 182–211 (1976)
60. Farr, W.H., Smith, O.D.: Statistical Modeling and Estimation of Reliability Functions for Software (SMERFS) Users Guide. Technical Report NAVSWC TR-84-373, Naval Surface Weapons Center (1993)
61. Fenton, N.: Software measurement: A necessary scientific basis. IEEE Trans. Softw. Eng. **20**(3), 199–206 (1994)
62. Fenton, N.E., Neil, M.: A critique of software defect prediction models. IEEE Trans. Softw. Eng. **25**(5), 675–689 (1999). doi:http://dx.doi.org/10.1109/32.815326
63. Festinger, L.: A Theory of Cognitive Dissonance. Stanford University Press, Stanford (1957)
64. Ficalora, J.P., Cohen, L.: Quality Function Deployment and Six Sigma. A QFD Handbook, 2nd edn. Prentice Hall, Englewood Cliffs (2010)
65. Florac, W.A., Carleton, A.D.: Measuring the Software Process: Statistical Process Control for Software Process Improvement. Addison-Wesley, Reading (1999)

66. Franch, X., Carvallo, J.P.: Using quality models in software package selection. IEEE Softw. **20**(1), 34–41 (2003)
67. Frank, M.: Konzeption und Einführung eines QM-Systems für Software. Diplomarbeit, Technische Universität München (2010)
68. Frankl, P., Hamlet, D., Littlewood, B., Strigini, L.: Choosing a testing method to deliver reliability. In: Proceedings of the 19th International Conference on Software Engineering (ICSE'97), pp. 68–78. ACM Press, New York (1997)
69. Frye, C.: CMM founder: Focus on the product to improve quality. http://searchsoftwarequality.techtarget.com/news/interview/0,289202,sid92_gci1316385,00.html (2008)
70. Gall, H., Jazayeri, M., Klösch, R., Trausmuth, G.: Software evolution observations based on product release history. In: Proceedings of the International Conference on Software Maintenance (ICSM'97), pp. 160–166. IEEE Computer Society, Silver Spring (1997)
71. Garvin, D.A.: What does "product quality" really mean? MIT Sloan Manag. Rev. **26**(1), 25–43 (1984)
72. Georgiadou, E.: GEQUAMO—a generic, multilayered, cusomisable, software quality model. Softw. Qual. J. **11**, 313–323 (2003)
73. Gilb, T., Graham, D.: Software Inspection. Addison-Wesley, Reading (1994)
74. Glass, R.: A classification system for testing, Part 2. IEEE Softw. **26**(1), 104 –104 (2009)
75. Gleirscher, M., Golubitskiy, D., Irlbeck, M., Wagner, S.: On the benefit of automated static analysis for small and medium-sized software enterprises. In: Proceedings of the Software Quality Days 2012. Lecture Notes in Business Information Processing, vol. 94, pp. 14–38 (2012)
76. Glinz, M.: Rethinking the notion of non-functional requirements. In: Proceedings of the Third World Congress for Software Quality, vol. II, pp. 55–64 (2005)
77. Grady, R.B., Caswell, D.L.: Software Metrics: Establishing a Company-Wide Program. Prentice Hall, Englewood Cliffs (1987)
78. Graham, D., Fewster, M.: Software Test Automation: Effective Use of Test Execution Tools, illustrated edn. Addison Wesley, Reading (1999)
79. Gruber, H., Plösch, R., Saft, M.: On the validity of benchmarking for evaluating code quality. In: Proceedings of the IWSM/MetriKon/Mensura 2010 (2010)
80. van Gurp, J., Bosch, J.: Design erosion: Problems and causes. J. Syst. Softw. **61**(2), 105–119 (2002)
81. Hayes, J.H., Zhao, L.: Maintainability prediction: A regression analysis of measures of evolving systems. In: Proceedings of the 21st IEEE International Conference on Software Maintenance (ICSM'05), pp. 601–604. IEEE Computer Society, Silver Spring (2005)
82. Heitlager, I., Kuipers, T., Visser, J.: A practical model for measuring maintainability. In: Proceedings of the 6th International Conference on Quality of Information and Communications Technology (2007)
83. Homeland Security: Common attack pattern enumeration and classification (CAPEC). Available Online at http://capec.mitre.org/. Accessed Oct 2008
84. Homeland Security: Common weakness enumeration (CWE). Available Online at http://cwe.mitre.org/. Accessed in Oct 2008
85. Huang, L., Boehm, B.: How much software quality investment is enough: A value-based approach. IEEE Softw. **23**(5), 88–95 (2006)
86. Hudepohl, J.P., Aud, S.J., Koshgoftaar, T.M., Allen, E.B., Mayrand, J.: Emerald: Software metrics and models on the desktop. IEEE Softw. **13**(5), 56–60 (1996)
87. IEEE: Standard 830-1998: Recommended practice for software requirements specifications (1998)
88. IEEE 1219: Software maintenance (1998)
89. ISO 15005:2002: Road vehicles – ergonomic aspects of transport information and control systems – dialogue management principles and compliance procedures (2002)
90. ISO 9000:2005: Quality management systems – fundamentals and vocabulary (2005)
91. ISO 9001:2008: Quality management systems – requirements (2008)

92. ISO/IEC 14598: Information technology – software product evaluation (1999)
93. ISO/IEC 15504-1:2004: Information technology – process assessment – Part 1: Concepts and vocabulary (2004)
94. ISO/IEC 15939:2007: Systems and software engineering – measurement process (2007)
95. ISO/IEC 25000:2005: Systems and software engineering – systems and software quality requirements and evaluation (SQuaRE) – guide to SQuaRE (2005)
96. ISO/IEC 25001:2007: Systems and software engineering – systems and software quality requirements and evaluation (SQuaRE) – planning and management (2007)
97. ISO/IEC 25010:2011: Systems and software engineering – systems and software quality requirements and evaluation (SQuaRE) – system and software quality models (2011)
98. ISO/IEC 25012:2008: Systems and software engineering – systems and software quality requirements and evaluation (SQuaRE) – data quality model (2008)
99. ISO/IEC 25020:2007: Systems and software engineering – systems and software quality requirements and evaluation (SQuaRE) – measurement reference model and guide (2007)
100. ISO/IEC 25021:2012: Systems and software engineering – systems and software quality requirements and evaluation (SQuaRE) – quality measure element (2012)
101. ISO/IEC 25030:2007: Systems and software engineering – systems and software quality requirements and evaluation (SQuaRE) – quality requirements (2007)
102. ISO/IEC 25040:2011: Systems and software engineering – systems and software quality requirements and evaluation (SQuaRE) – evaluation process (2011)
103. ISO/IEC 25041:2012: Systems and software engineering – systems and software quality requirements and evaluation (SQuaRE) – evaluation guide for developers, acquirers and independent evaluators (2012)
104. ISO/IEC 25045:2010: Systems and software engineering – systems and software quality requirements and evaluation (SQuaRE) – evaluation module for recoverability (2010)
105. ISO/IEC 26262:2011: Road vehicles – functional safety (2011)
106. ISO/IEC 27001: Information technology – security techniques – information security management systems – requirements (2005)
107. ISO/IEC TR 9126-1:2001: Software engineering – product quality – Part 1: Quality model (2001)
108. ISO/IEC/IEEE 24765:2010: Systems and software engineering – vocabulary (2010)
109. Jones, C.: Applied Software Measurement: Assuring Productivity and Quality. McGraw-Hill, New York (1991)
110. Jones, C.: Software Assessments, Benchmarks, and Best Practices. Addison-Wesley Longman Publishing Co., Boston (2000)
111. Jones, W.D., Vouk, M.A.: Field Data Analysis. In: Lyu, M.R. (ed.) Handbook of Software Reliability Engineering, Chap. 11. IEEE Computer Society Press/McGraw-Hill, Silver Spring/New York (1996)
112. Juergens, E., Deissenboeck, F., Feilkas, M., Hummel, B., Schaetz, B., Wagner, S., Domann, C., Streit, J.: Can clone detection support quality assessments of requirements specifications? In: Proceedings of the 32nd ACM/IEEE International Conference on Software Engineering (ICSE'10), pp. 79–88. ACM Press, New York (2010)
113. Juergens, E., Deissenboeck, F., Hummel, B., Wagner, S.: Do code clones matter? In: Proceedings of the International Conference on Software Engineering (ICSE'09). IEEE Computer Society, Silver Spring (2009)
114. Juran, J.M.: Juran's Quality Control Handbook. McGraw-Hill, New York (1988)
115. Juristo, N., Moreno, A.M., Vegas, S.: Reviewing 25 years of testing technique experiments. Empir. Softw. Eng. **9**, 7–44 (2004)
116. Juristo, N., Moreno, A.M., Vegas, S., Solari, M.: In search of what we experimentally know about unit testing. IEEE Softw. **23**(6), 72–80 (2006)
117. Kafura, D., Reddy, G.R.: The use of software complexity metrics in software maintenance. IEEE Trans. Softw. Eng. **13**(3), 335–343 (1987)
118. Kapser, C., Godfrey, M.W.: "Cloning considered harmful" considered harmful. In: Proceedings of the 13th Working Conference on Reverse Engineering (WCRE '06), pp. 19–28. IEEE Computer Society, Silver Spring (2006)

119. Khaddaj, S., Horgan, G.: A proposed adaptable quality model for software quality assurance. J. Comput. Sci. **1**(4), 482–487 (2005)
120. Kitchenham, B.: Towards a constructive quality model. Part I: Software quality modelling, measurement and prediction. Softw. Eng. J. **2**(4), 105–113 (1987)
121. Kitchenham, B., Linkman, S., Pasquini, A., Nanni, V.: The SQUID approach to defining a quality model. Softw. Qual. J. **6**(3), 211–233 (1997)
122. Kitchenham, B., Pfleeger, S.L.: Software quality: The elusive target. IEEE Softw. **13**(1), 12–21 (1996)
123. Kitchenham, B., Pfleeger, S.L., Fenton, N.: Towards a framework for software measurement validation. IEEE Trans. Softw. Eng. **21**(12), 929–944 (1995). doi:http://dx.doi.org/10.1109/32.489070
124. Kitchenham, B., Pickard, L.M.: Towards a constructive quality model. Part 2: Statistical techniques for modelling software quality in the ESPRIT REQUEST project. Softw. Eng. J. **2**(4), 114–126 (1987)
125. Kläs, M., Heidrich, J., Münch, J., Trendowicz, A.: CQML scheme: A classification scheme for comprehensive quality model landscapes. In: Proceedings of the 35th Euromicro Conference on Software Engineering and Advanced Applications (2009)
126. Knox, S.T.: Modeling the cost of software quality. Digit. Technol. J. **5**(4), 9–17 (1993)
127. Kof, L.: An application of natural language processing to domain modelling – two case studies. Int. J. Comput. Syst. Sci. Eng. **20**, 37–52 (2005)
128. Koschke, R.: Survey of research on software clones. In: Duplication, Redundancy, and Similarity in Software. Dagstuhl Seminar Proceedings (2007)
129. Kotter, J.: Leading Change. Harvard Business School Press, Boston (1996)
130. Kotter, J., Cohen, D.: The Heart of Change: Real-Life Stories of How People Change Their Organizations. Harvard Business School Press, Boston (2002)
131. Krasner, H.: Using the cost of quality approach for software. Crosstalk **11**, 6–11 (1998)
132. Laitenberger, O.: A Survey of Software Inspection Technologies. In: Handbook on Software Engineering and Knowledge Engineering, vol. 2, pp. 517–555. World Scientific, Singapore (2002)
133. Larkin, T., Larkin, S.: Communicating Change: How to Win Employee Support for New Business Directions. McGraw-Hill, New York (1994)
134. Lehman, M.M.: On understanding laws, evolution, and conservation in the large-program life cycle. J. Syst. Softw. **1**, 213–221 (1980)
135. Lindvall, M., Donzelli, P., Asgari, S., Basili, V.: Towards reusable measurement patterns. In: Proceedings of the 11th IEEE International Software Metrics Symposium (METRICS'05). IEEE Computer Society, Silver Spring (2005)
136. Littlewood, B., Verall, J.: A Bayesian Reliability Growth Model for Computer Software. Appl. Stat. **22**(3), 332–346 (1973)
137. Lochmann, K.: Engineering quality requirements using quality models. In: Proceedings of the 15th IEEE International Conference on Engineering of Complex Computer Systems. IEEE Computer Society, Silver Spring (2010)
138. Luckey, M., Baumann, A., Méndez Fernández, D., Wagner, S.: Reusing security requirements using an extend quality model. In: Proceedings of the 2010 ICSE Workshop on Software Engineering for Secure Systems (2010)
139. Lyu, M.R. (ed.): Handbook of Software Reliability Engineering. IEEE Computer Society Press/McGraw-Hill, Silver Spring/New York (1996)
140. MacKay, D.J.C.: Information Theory, Inference, and Learning Algorithms. Cambridge Press, Cambridge (2003)
141. Mandeville, W.: Software costs of quality. IEEE J. Sel. Areas Commun. **8**(2), 315–318 (1990)
142. Marinescu, C., Marinescu, R., Mihancea, R.F., Ratiu, D., Wettel, R.: iPlasma: An integrated platform for quality assessment of object-oriented design. In: Proceedings of the 21st IEEE International Conference on Software Maintenance. IEEE Computer Society, Silver Spring (2005)

143. Marinescu, R., Ratiu, D.: Quantifying the quality of object-oriented design: The factor-strategy model. In: Proceedings of the 11th Working Conference on Reverse Engineering (WCRE'04), pp. 192–201. IEEE Computer Society, Silver Spring (2004)
144. The MathWorks: Simulink Reference (2006)
145. MathWorks Automotive Advisory Board: Controller style guidelines for production intent using Matlab, Simulink and Stateflow. http://www.mathworks.com/industries/auto/maab.html (2001)
146. Mayr, A., Plösch, R., Kläs, M., Lampasona, C., Saft, M.: A comprehensive code-based quality model for embedded systems. In: Proceedings of the 23rd IEEE International Symposium on Software Reliability Engineering (ISSRE 2012). IEEE Computer Society, Silver Spring (2012)
147. McCabe, T.: A complexity measure. IEEE Trans. Softw. Eng. **SE-2**(4), 308–320 (1976)
148. McCall, J., Walters, G.: Factors in Software Quality. The National Technical Information Service, Springfield (1977)
149. McCall, J.A., Richards, P.K., Walters, G.F.: Factors in Software Quality. National Technical Information Service, Springfield (1977)
150. McFeeley, B.: IDEAL: A user's guide for software process improvement. In: Handbook CMU/SEI-96-HB-001. Software Engineering Institute, Pittsburgh (1996)
151. Menzies, T., Butcher, A., Marcus, A., Zimmermann, T., Cok, D.: Loval vs. global models for effort estimation and defect prediction. In: Proceedings of the 26th IEEE/ACM International Conference on Automated Software Engineering (ASE '11), pp. 343–351. IEEE Computer Society, Silver Spring (2011)
152. MISRA AC SLSF: Modelling design and style guidelines for the application of simulink and stateflow (2009)
153. MISRA AC TL: Modelling style guidelines for the application of targetlink in the context of automatic code generation (2007)
154. Monden, Y.: Toyota Production System. An Integrated Approach to Just-In-Time, 3rd edn. Engineering & Management Press (1998)
155. Mordal-Manet, K., Balmas, F., Denier, S., Ducasse, S., Wertz, H., Laval, J., Bellingard, F., Vaillergues, P.: The squale model – a practice-based industrial quality model. In: Proceedings of the IEEE International Conference on Software Maintenance (2009)
156. Münch, J., Kläs, M.: Balancing upfront definition and customization of quality models. In: Workshop-Band Software-Qualitätsmodellierung und -bewertung (SQMB 2008). Technische Universität München (2008)
157. Musa, J., Ackerman, A.: Quantifying software validation: When to stop testing? IEEE Softw. **6**(3), 19–27 (1989)
158. Musa, J.D.: Software Reliability Engineering: More Reliable Software Faster and Cheaper, 2nd edn. AuthorHouse, Bloomington (2004)
159. Musa, J.D., Iannino, A., Okumoto, K.: Software Reliability: Measurement, Prediction, Application. McGraw-Hill, New York (1987)
160. Musa, J.D., Okumoto, K.: A logarithmic poisson execution time model for software reliability measurement. In: Proceedings of the Seventh International Conference on Software Engineering (ICSE'84), pp. 230–238. ACM Press, New York (1984)
161. Myers, G.J.: The Art of Software Testing. Wiley, New York (1979)
162. Nagappan, N., Ball, T.: Static analysis tools as early indicators of pre-release defect density. In: Proceedings of the International Conference on Software Engineering (ICSE '05). ACM Press, New York (2005)
163. Nagappan, N., Ball, T., Zeller, A.: Mining metrics to predict component failures. In: Proceedings of the 28th International Conference on Software Engineering (ICSE'06), pp. 452–461. ACM Press, New York (2006)
164. Nagel, P.M., Scholz, F.W., Skrivan, J.A.: Software Reliability: Additional Investigations into Modeling with Replicated Experiments. NASA Contractor Rep. 172378, NASA Langley Res. Center (1984)

165. Nagel, P.M., Skrivan, J.A.: Software Reliability: Repetitive Run Experimentation and Modeling. NASA Contractor Rep. 165836, NASA Langley Res. Center (1982)
166. Neuhaus, S., Zimmermann, T., Holler, C., Zeller, A.: Predicting vulnerable software components. In: Proceedings of the 14th ACM Conference on Computer and Communications Security (CCS '07), pp. 529–540. ACM Press, New York (2007)
167. Nuseibeh, B., Easterbrook, S.: Requirements engineering: A roadmap. In: Proceedings of the Conference on the Future of Software Engineering (ICSE '00), pp. 35–46. ACM Press, New York (2000)
168. Oman, P., Hagemeister, J.: Metrics for assessing a software system's maintainability. In: Proceedings of the International Conference on Software Maintenance (1992)
169. Ortega, M., Pérez, M., Rojas, T.: Construction of a systemic quality model for evaluating a software product. Softw. Qual. J. **11**, 219–242 (2003)
170. Parnas, D.L.: Software aging. In: Proceedings of the International Conference on Software Engineering (ICSE '94), pp. 279–287. IEEE Computer Society, Silver Spring (1994)
171. Pfleeger, S.L., Atlee, J.M.: Software Engineering: Theory and Practice, 4th edn. Prentice Hall, Englewood Cliffs (2009)
172. Pham, H.: Software Reliability. Springer, New York (2000)
173. Plato, (Translator), R.W.: Symposium, reprint edn. Oxford World's Classics. Oxford University Press, Oxford (1998)
174. Plösch, R., Gruber, H., Hentschel, A., Körner, C., Pomberger, G., Schiffer, S., Saft, M., Storck, S.: The EMISQ method and its tool support – expert based evaluation of internal software quality. J. Innov. Syst. Softw. Eng. **4**(1) (2008)
175. Plösch, R., Gruber, H., Körner, C., Pomberger, G., Schiffer, S.: A proposal for a quality model based on a technical topic classification. In: Tagungsband des 2. Workshops zur Software-Qualitätsmodellierung und -bewertung (2009)
176. Plösch, R., Gruber, H., Pomberger, G., Saft, M., Schiffer, S.: Tool support for expert-centred code assessments. In: Proceedings of the International Conference on Software Testing, Verification, and Validation (ICST), pp. 258–267. IEEE Computer Society, Silver Spring (2008)
177. Plösch, R., Mayr, A., Körner, C.: Collecting quality requirements using quality models and goals. In: Proceedings of the 2010 Seventh International Conference on the Quality of Information and Communications Technology (2010)
178. Pohl, K., Rupp, C.: Requirements Engineering Fundamentals. Rocky Nook, Santa Barbara (2011)
179. Poppendieck, M., Poppendieck, T.: Lean Software Development. An Agile Toolkit. Addison-Wesley Professional, Reading (2003)
180. Pretschner, A., Prenninger, W., Wagner, S., Kühnel, C., Baumgartner, M., Sostawa, B., Zölch, R., Stauner, T.: One evaluation of model-based testing and its automation. In: Proceedings of the 27th International Conference on Software Engineering (ICSE'05). ACM Press, New York (2005)
181. Puchner, S.: Sustainable change in organizations. In: Wagner, S., Deissenboeck, F., Hummel, B., Juergens, E., y Parareda, B.M., Schaetz, B. (eds.) Selected Topics in Software Quality. Technische Universität München, Garching (2008)
182. Reel, J.S.: Critical success factors in software projects. IEEE Softw. **16**(3), 18–23 (1999)
183. Robertson, S., Robertson, J.: Mastering the Requirements Process. ACM Press/Addison-Wesley, New York/Reading (1999)
184. Runeson, P., Andersson, C., Thelin, T., Andrews, A., Berling, T.: What do we know about defect detection methods? IEEE Softw. **23**(3), 82–90 (2006)
185. Samoladas, I., Gousios, G., Spinellis, D., Stamelos, I.: The SQO-OSS quality model: Measurement based open source software evaluation. In: Proceedings of the 4th International Conference on Open Source Systems, vol. 275. Springer, New York (2008)
186. Sanchez, J., Williams, L., Maximilien, E.: On the sustained use of a test-driven development practice at IBM. In: Proceedings of the AGILE 2007, pp. 5–14. IEEE Computer Society, Silver Spring (2007)

187. Schackmann, H., Jansen, M., Lichter, H.: Tool support for user-defined quality assessment models. In: Proceedings of the MetriKon 2009 (2009)
188. Shewhart, W.A.: Statistical Method from the Viewpoint of Quality Control. Dover, New York (1986)
189. Shull, F., Rus, I., Basili, V.: How perspective-based reading can improve requirements inspections. Computer **33**(7), 73–79 (2000)
190. Slaughter, S.A., Harter, D.E., Krishnan, M.S.: Evaluating the cost of software quality. Commun. ACM **41**(8), 67–73 (1998)
191. van Solingen, R., Berghout, E.: Goal/Question/Metric Method. McGraw-Hill Professional, New York (1999)
192. Sommerville, I.: Software Engineering, 9th edn. Addison Wesley, Reading (2010)
193. Spillner, A., Linz, T., Roßner, T., Winter, M.: The Software Testing Practice: Test Management: A Study Guide for the Certified Tester Exam ISTQB Advanced Level. Rocky Nook, Santa Barbara (2007)
194. Sun Microsystems: Secure coding guidelines for the java programming language, version 2.0. Available Online at http://java.sun.com/security/seccodeguide.html
195. Tenhunen, V., Sajaniemi, J.: An evaluation of inspection automation tools. In: Proceedings of the European Conference on Software Quality (ECSQ'02). Lecture Notes in Computer Science, vol. 2349, pp. 351–361. Springer, Berlin (2002)
196. Thelin, T., Runeson, P., Wohlin, C.: Prioritized use cases as a vehicle for software inspections. IEEE Softw. **20**(4), 30–33 (2003)
197. Tian, J.: Quality-Evaluation Models and Measurements. IEEE Softw. **21**(3), 84–91 (2004)
198. Tian, J.: Software Quality Engineering. Testing, Quality Assurance, and Quantifiable Improvement. Wiley, New York (2005)
199. Voas, J.: Can clean pipes produce dirty water? IEEE Softw. **14**(4), 93–95 (1997)
200. Wagner, S.: A literature survey of the quality economics of defect-detection techniques. In: Proceedings of the 5th ACM-IEEE International Symposium on Empirical Software Engineering (ISESE'06), pp. 194–203. ACM Press, New York (2006)
201. Wagner, S.: Using economics as basis for modelling and evaluating software quality. In: Proceedings of the First International Workshop on the Economics of Software and Computation (ESC-1) (2007)
202. Wagner, S.: Cost-Optimisation of Analytical Software Quality Assurance. VDM Verlag Dr. Müller, Saarbrücken (2008)
203. Wagner, S.: A Bayesian network approach to assess and predict software quality using activity-based quality models. Inf. Softw. Technol. **52**(11), 1230–1241 (2010)
204. Wagner, S., Deissenboeck, F.: An integrated approach to quality modelling. In: Proceedings of the 5th Workshop on Software Quality (5-WoSQ). IEEE Computer Society (2007)
205. Wagner, S., Deissenboeck, F., Aichner, M., Wimmer, J., Schwalb, M.: An evaluation of two bug pattern tools for java. In: Proceedings of the International Conference on Software Testing, Verification and Validation (ICST'08). IEEE Computer Society, Silver Spring (2008)
206. Wagner, S., Fischer, H.: A Software Reliability Model Based on a Geometric Sequence of Failure Rates. Technical Report TUMI-0520, Institut für Informatik, Technische Universität München (2005)
207. Wagner, S., Fischer, H.: A Software Reliability Model Based on a Geometric Sequence of Failure Rates. In: Proceedings of the 11th International Conference on Reliable Software Technologies (Ada-Europe '06). Lecture Notes in Computer Science, vol. 4006, pp. 143–154. Springer, Berlin (2006)
208. Wagner, S., Jürjens, J.: Model-based identification of fault-prone components. In: Proceedings of the Fifth European Dependable Computing Conference (EDCC-5). Lecture Notes in Computer Science, vol. 3463, pp. 435–452. Springer, New York (2005)
209. Wagner, S., Jürjens, J., Koller, C., Trischberger, P.: Comparing bug finding tools with reviews and tests. In: Proceedings of the 17th International Conference on Testing of Communicating Systems (TestCom'05). Lecture Notes in Computer Science, vol. 3502, pp. 40–55. Springer, New York (2005)

210. Wagner, S., Lochmann, K., Heinemann, L., Kläs, M., Lampasona, C., Trendowicz, A., Plösch, R., Mayr, A., Seidl, A., Goeb, A., Streit, J.: Practical product quality modelling and assessment: The Quamoco approach (in preparation)
211. Wagner, S., Lochmann, K., Heinemann, L., Kläs, M., Trendowicz, A., Plösch, R., Seidl, A., Goeb, A., Streit, J.: The Quamoco product quality modelling and assessment approach. In: Proceedings of the 34th International Conference on Software Engineering. IEEE Computer Society, Silver Spring (2012)
212. Wagner, S., Lochmann, K., Winter, S., Goeb, A., Klaes, M., Nunnenmacher, S.: Quality models in practice: Survey results. https://quamoco.in.tum.de/wordpress/wp-content/uploads/2010/01/Software_Quality_Models_in_Practice.pdf (2010)
213. Wagner, S., Lochmann, K., Winter, S., Goeb, A., Kläs, M., Nunnenmacher, S.: Software quality in practice. survey results. Technical Report TUM-I128, Technische Universität München (2012)
214. Wagner, S., Méndez Fernández, D., Islam, S., Lochmann, K.: A security requirements approach for web systems. In: Proceedings of the Quality Assessment in Web (QAW 2009). CEUR (2009)
215. Wallace, D.R., Fujii Roger, U.: Software verification and validation: An overview. IEEE Softw. **6**(3), 10–17 (1989)
216. Wiesmann, A., van der Stock, A., Curphey, M., Stirbei, R. (eds.): A Guide to Building Secure Web Applications and Web Services. OWASP (2005)
217. Winter, S., Wagner, S., Deissenboeck, F.: A comprehensive model of usability. In: Proceedings of the Engineering Interactive Systems 2007. Lecture Notes in Computer Science, vol. 4940, pp. 106–122. Springer, New York (2008)
218. Womack, J.P., Jones, D.T., Roos, D.: The Machine That Changed the World: The Story of Lean Production, reprint edn. HarperPaperbacks, Hammersmith (1991)
219. van Zeist, R.H.J., Hendriks, P.R.H.: Specifying software quality with the extended ISO model. Softw. Qual. J. **5**(4), 273–284 (1996)
220. Zheng, J., Williams, L., Nagappan, N., Snipes, W., Hudepohl, J.P., Vouk, M.A.: On the value of static analysis for fault detection in software. IEEE Trans. Softw. Eng. **32**(4), 240–253 (2006)
221. Zhou, Y., Leung, H.: Predicting object-oriented software maintainability using multivariate adaptive regression splines. J. Syst. Softw. **80**(8), 1349–1361 (2007)

# Index



S. Wagner, *Software Product Quality Control*, DOI 10.1007/978-3-642-38571-1,